Reproduction of
Eukaryotic Cells

Reproduction of Eukaryotic Cells

David M. Prescott

Department of Molecular, Cellular
and Developmental Biology
University of Colorado
Boulder, Colorado

ACADEMIC PRESS   New York   San Francisco   London   1976

A Subsidiary of Harcourt Brace Jovanovich, Publishers

ACADEMIC PRESS, INC.
111 Fifth Avenue, New York, New York 10003

United Kingdom Edition published by
ACADEMIC PRESS, INC. (LONDON) LTD.
24/28 Oval Road, London NW1

Library of Congress Cataloging in Publication Data

Prescott, David M
    Reproduction of eukaryotic cells.

    Bibliography: p.
    Includes index.
    1.   Cell cycle.   2.   Cell proliferation.
I.   Title.   [DNLM: 1.   Cell division.   QH605
P929r]
QH605.P7   574.8'762   75-36654
ISBN 0–12–564150–8

to Gayle

Contents

PREFACE   ix

1 Introduction

Sections of the Cell Cycle   2
Regulation of Cell Reproduction   4

2 Cell Growth through the Cycle

Fission Yeast (*Schizosaccharomyces pombe*)   8
Amoeba proteus   10
Tetrahymena pyriformis   14
Mouse Fibroblast   16
Conclusions   17

3 Cell Synchrony

Natural Synchrony   19
Experimentally Derived Synchrony   21

4 The G_1 Period

Variableness of G_1 36
Relation of Cell Growth to the Length of G_1 39
Control of Cell Reproduction in G_1 44
Requirements for Protein and RNA Synthesis to Complete G_1 52
Preparation for DNA Synthesis   55

5 Initiation of the S Period

Role of Nuclear–Cytoplasmic Interactions in DNA Synthesis   60
Intranuclear Site of Initiation and Continuation of DNA Synthesis   67

6 The S Period

The Number of Replicating Units (Replicons)   74
Minimum Number of Replicon Families   75
Apparent Change in the Number of Origins of Replication in
  Relation to Development   76

The Ordering of DNA Replication   78
Requirements for Protein and RNA Synthesis during the S Period   83
The Length of the S Period in Cells of Different Ploidies   85

7 The G_2 Period

Preparation for Mitosis   87
Arrest of the Cycle in G_2 88
Requirements for Protein and RNA Syntheses during G_2 89
The G_2 to D Transition   90

8 Activities during Cell Division

RNA Synthesis during Mitosis   91
Protein Synthesis during Mitosis   93
Reversible Shift of Nuclear RNA and Protein to the Cytoplasm during Mitosis   94

9 Cell Surface Changes during the Cycle

Morphological Changes   97
Chemical Changes of the Plasma Membrane   103
Plant Lectin Binding and Cell Agglutinability   104

10 Cyclic AMP, Cyclic GMP, and Cell Reproduction

cAMP and the Cell Cycle   107
Effects of Experimentally Induced Increases in Intracellular cAMP on Cell Growth   112
Cyclic Nucleotides and the Imposition and Release of G_1 Arrest   113
Regulation of cAMP Levels in Cells   117

11 Nuclear Proteins and the Cell Cycle

Histones   119
Nonhistone Nuclear Proteins   122

12 RNA Synthesis and the Cell Cycle

Text   124

13 Patterns of Enzyme Activities through the Cell Cycle

Text   127

14 The Genetics of the Cell Cycle

How Much of the Genome Is Specifically Concerned with the Cell Cycle?   130
Genetic Analysis of the Cell Cycle in Yeast   131
Genetic Analysis of the Cell Cycle in Mammalian Cells   133
The Concept of Regulatory Genes for Cell Reproduction   135
Final Comment   139

Bibliography   141

INDEX   169

Preface

The purpose of this book is to organize in a single source the principal facts and observations on the cell life cycle and reproduction of eukaryotic cells in an effort to increase our overall understanding of how these cells reproduce themselves and how this reproduction is regulated. Obviously, these are matters of widespread importance in the biological and biomedical sciences.

In some parts, this book overlaps with J. M. Mitchison's book on "Biology of the Cell Cycle" (Cambridge University Press, 1971), but most of the areas covered in Mitchison's book are omitted here or only dealt with briefly. It is my intention that the two books be complementary. This is reflected in the fact that most of the 563 papers listed in the references have been published since the appearance of Mitchison's book.

I acknowledge with thanks the helpful reviews by my colleagues Gretchen Stein and Lawrence Allred. I am most grateful to Gayle Prescott for her patience, skill, and perseverance in preparation of the manuscript for publication.

David M. Prescott

1
Introduction

The ability to reproduce is so fundamental to cell existence that it is properly considered as a major defining property of the cell. Every cell comes into being through reproduction and depends upon cell reproduction for its long-term survival. Cells may survive for extended periods without reproduction when they develop into specialized forms, such as the spores and cysts of unicellular organisms, the quiescent cells of plant embryos in seeds, or the differentiated cellular forms in multicellular organisms generally; but cells that relinquish or lose the ability to reproduce or are deprived too long of the possibility for reproduction will die. In sum, any cell that does not reproduce has a limited future.

The continuation and propagation of every species of organism obviously depends directly on cell reproduction. Among unicellular organisms each cell reproduction increases the species by one member. Among multicellular organisms cell reproduction provides for the continuity of the germ line of a species and provides the somatic cells required to build and maintain individuals. Cell proliferation is therefore a familiar component of the development of every multicellular organism, although the magnitude of cell reproduction in adult organisms is frequently not appreciated. Adult humans consist of about 100 trillion cells (10^{14} cells), all derived from a single cell, the fertilized ovum, through the process of cell reproduction. In the adult a large amount of continuous cell reproduction is essential for replacement of cells that die or are otherwise lost. An adult human contains about 2.5×10^{13} erythrocytes (5 liters of blood/body with 5×10^6 erythrocytes/mm^3), and the average life time of an erythrocyte is 120 days (10^7 seconds). Therefore, to maintain the erythrocyte population, precursor cells of the erythrocyte must produce 2.5×10^{13} new cells every 10^7 seconds, which is equivalent to a continuously sustained rate of 2.5×10^6 cell divisions per second. Similarly, the trillion lymphocytes in an adult are replaced at a rate of 2×10^5 per second. The total rate of division for all renewing cell populations (skin, intestinal epithelium, leukocytes, etc.) probably exceeds 20×10^6 divisions per second. This amount of cell reproduction is impressive, but is also remarkable for the precision with which it is regulated. The rates of cell reproduction vary from tissue to tissue, but in each tissue, production of new cells exactly balances the loss of cells. How regulation of cell reproduction is achieved and how radiation, oncogenic viruses, and chemical mutagens (carcinogens) cause loss of regulation are topics that occupy a major place in contemporary research in cell biology.

Cell reproduction consists of three components: growth, DNA replication, and cell division. During each life cycle a cell grows by doubling all of its structural elements and functional capacities. DNA replication is, of course, a part of growth, but is singularly important both because it is an absolute, genetic prerequisite for successful cell division and because it is the key event around which the step by step progress of a cell through its life cycle is arranged. All the events that make up cell growth and chromosome replication are integrated with one another to bring about the orderly progression of the cell cycle, culminating in the precise distribution of daughter chromosomes to form daughter nuclei and the splitting of the cell into two daughters.

Detailed knowledge of the molecular mechanisms responsible for the multitude of integrative interactions that coordinate cell growth, DNA replication, and cell division will be required if we are to understand finally the complex process of cell reproduction and its regulation. Insight into these problems is beginning to emerge from the wealth of accumulated observations on events in the life cycles of many different kinds of cells. The purpose of this book is to review how far the accumulated facts have led us toward an understanding of cell reproduction. Thus, one overall, primary aim of the discussions to follow is assessment of the extent to which we can now begin to define a unified set of principles that governs the operation of the cycle for cells in general.

SECTIONS OF THE CELL CYCLE

The cell cycle is ordinarily considered to begin with the completion of one cell division and to end with the completion of the next division (Fig. 1), and the time taken for one cell cycle is the *generation time*. Cell division is a convenient marker because it can be so readily observed or measured, but in a strict sense the beginning and end of the cell cycle is that point in interphase, usually early interphase, at which the decision is made to stop proliferation or to proceed to the next cell division (see Chapters 4 and 14).

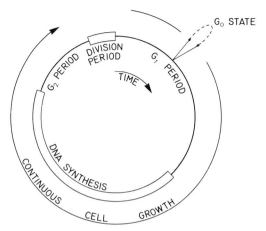

Fig. 1. The major features of the cell life cycle. The relative proportions of the cycle may vary considerably from one kind of cell to another, but the reproduction of every cell consists of growth coupled with DNA replication followed by cell division. A mammalian cell growing in culture with a generation time of 16 hours, for example, will have a $G_1 = 5$ hours, S = 7 hours, $G_2 = 3$ hours, and D = 1 hour. G_0 is the state into which cells are postulated to move when the cell cycle is arrested in G_1 by various kinds of environmental conditions.

Progress through the cycle is usually assessed by observing two readily identified processes, DNA replication and cell division. As first shown with plant root cells by Howard and Pelc (237), these two steps allow the cycle to be divided into four successive intervals, G_1, S, G_2, and D (Fig. 1). G_1 is the time gap between the completion of cell division and the beginning of DNA replication; S is the period of DNA replication; G_2 is the time gap between the end of DNA replication and the onset of cell division; and D is the time taken for cell division. The D period is also sometimes called the M period (for mitosis).

G_1, S, and G_2 are periods of continuous cell growth with general increases in all the cell's structures and functional capacities. During the division period, at least in mitotically dividing cells, the rate of growth falls sharply (Chapter 8). With the completion of mitosis the growth rate rises quickly.

The S and D periods are defined, respectively, by DNA replication and cell division, but no specific events have been similarly identified that can account for progress of the cell through the G_1 and G_2 periods. Hence, G_1 and G_2 represent major gaps in our understanding of the cause and effect continuity of the cell life cycle. Although specific events have not been identified, the G_1 period is generally assumed to contain a succession of events that leads to the initiation of DNA replication; unfortunately we still know virtually nothing of what these events might be. Information about the molecular basis of G_1 is particularly crucial because regulation of cell reproduction usually consists of the control of cell transit through this part of the cycle. The G_2 period is believed to reflect a sequence of steps initiated by the completion of DNA synthesis and leading to nuclear division and cytokinesis, but the specific facts are few. In addition to identifying those events responsible for progression through the G_1 and G_2 periods, much still remains to be done to complete our understanding of the molecular events that make up the orderly transit of the cell through DNA replication and cell division. The properties of the four sections are discussed in detail in Chapters 4–8.

REGULATION OF CELL REPRODUCTION

The matter of the regulation of cell production is briefly introduced here because it is so closely tied to the study of the progression of the cell cycle. More extensive discussion of cycle regulation, particularly the genetic basis of regulation, is included in Chapter 14 after a more detailed description of the composition of the cell cycle.

Regulation is achieved by interrupting the progress of a cell through its cycle. It is therefore evident that a detailed knowledge of how the cell cycle works is a prerequisite for learning how cell reproduction is regulated. The molecular mechanisms that effect this interruption of the cycle are still poorly understood although it is clear that they work by stopping the cell at a point in G_1 and sometimes in the G_2 period, but never in S or D.

Nonspecific vs Specific Regulation

Among unicellular organisms regulation of cell reproduction consists of stopping the cycle when environmental conditions become unfavorable for further cell reproduction, for example, by exhaustion of some essential nutrient. The cells of multicellular organisms growing in culture respond in similar fashion to relatively nonspecific conditions such as deprivation of one or another required growth factor (see 232, 233, 233a, 367). Presumably, cells sense that some environmental condition has become unfavorable for growth because it affects cell metabolism adversely, but how the cell translates an adverse effect on metabolism into G_1 arrest of the cycle is not known.

Within multicellular organisms, in contrast to what occurs in unicellular organisms or cell cultures, the regulation of cell reproduction is not normally accomplished by regulation of the supply of nutrients; within multicellular organisms the cellular environment is maintained in a relatively stable state that is favorable for cell reproduction, yet cells do not reproduce freely. Regulation of reproduction is a highly specific process that is achieved by hormones and by hormonelike molecules called chalones that are secreted by cells. Some chalones work in an intratissue manner; the particular cell type producing the chalone is itself the specific target. In other cases, chalones act in an intertissue manner; the chalone produced by one cell type affects the reproduction of a different cell type or types. The study of chalones has been difficult, possibly because they are extremely labile molecules. The limited evidence now available indicates that chalones are proteins or glycoproteins.

Finally, the specific regulation of the various cell types by chalone-mediated, cell–cell interactions in multicellular organisms is somehow achieved by the arrest of the cell cycle at a point in G_1 and occasionally at a point in G_2. These G_1 and G_2 arrest points probably represent on–off switches operated by regulatory genes.

In summary, it is possible to identify two major types of regulation of reproduction, a relatively nonspecific type of regulation among unicellular organisms and a highly specific regulatory system among the cells of multicellular organisms. Both types of regulation operate by arresting the cell cycle in the G_1 period and occasionally in G_2. It is therefore possible that both nonspecific and specific regulatory signals ultimately impinge on the same cell cycle event in the G_1 period.

Specific Regulation of the Cell Cycle in Yeast

A phenomenon that occupies a position somewhere between the nonspecific regulation in unicellular eukaryotes or in cultured cells in response to environmental conditions and the specific regulation among cells in multicellular organisms occurs in the budding yeast *Saccharomyces cerevisiae*. Sexual mating occurs between different mating types, designated "α" and "a," which are determined by a single genetic locus. Each mating type produces and secretes a different mating factor (mating factor α and mating factor a) that blocks the cell cycle of the opposite mating type at a specific point in G_1 (Fig. 2) (78). This arrest of cells in G_1 is presumed to be necessary for successful mating between the two types (215, 534). The mating factor α is a small protein with a molecular weight of about 12,000 daltons (123). The mating factor a has not yet been characterized. The mating factors appear to arrest the cycle near or at the *same* point in G_1 as does cycle arrest caused by nutrient deprivation, suggesting that both kinds of regulation impinge on the same event in the G_1 progression (214). In short, a mechanism consisting of specific signals and specific sensitivity in the target cells has evolved for mutual regulation of the cell cycles between mating types in this unicellular eukaryote. The regulation of the cell cycle in budding yeast by mating factors is thus similar to the highly specific regulation of cell reproduction by chalone-mediated, cell–cell interaction in multicellular organisms. The yeast situation suggests that perhaps the principle of specific cell–cell interactions that is employed in control of cell reproduction in multi-

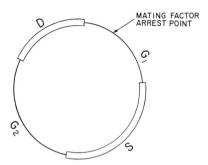

Fig. 2. The cell cycle in budding yeast. Mating types a and α produce mating factors that mediate the reciprocal arrest of the two mating strains in G_1. Conjugation takes place between cells of opposite mating types in G_1 arrest.

cellular organisms might have evolved first in unicellular organisms, to be subsequently exploited in the origination of the first multicellular organism.

The Concept of the G_0 State

The discovery that regulation of the cell cycle occurs primarily by G_1 arrest has led to the introduction of the term G_0 (279) to describe the state of the cell in the arrested condition (Fig.1). In the G_0 state the cell may be considered to have been withdrawn from the cell cycle. The withdrawal into G_0 is, for most cell types, reversible, and the cell may reenter the G_1 period of the cycle and resume proliferation. A few differentiated cell types in multicellular animals, primarily neurons and striated muscle cells, are irreversibly arrested in the G_0 state. Evidence for the existence of the G_0 state in cultured animal cells is discussed in Chapter 4.

In summary, the outlines of the cell cycle have been established, and this has led to a clearer delineation of the processes that make up cell reproduction and to a sharper view of how cell reproduction is regulated. Major tasks are now to determine the basic causal sequence of molecular events that underlies the procession of G_1-S-G_2-D, to explain how the many various growth activities of the cell participate in or are coordinated with this procession, and to discover the molecular nature of the mechanism by which the cycle is interrupted in the regulation of cell proliferation.

2
Cell Growth through the Cycle

Early efforts to understand cell reproduction, beginning around the turn of the century, were directed primarily toward the relationship between cell growth and cell division. Since, on the average, a cell doubles its size before it divides, it was reasonable to suppose that growth must somehow be a prerequisite for division. This led to the idea that the completion of a doubling in cell size, or at least growth to some particular cell size, somehow provided the trigger for the initiation of cell division. This general idea was developed mostly by Hertwig (227), who proposed that the nucleus is capable of supporting some maximum amount of cytoplasm. When this amount of cytoplasm was produced by growth, that is, when a particular cytoplasmic–nuclear ratio was attained, the cell was believed to enter an unstable state that somehow triggered cell division. The proposed causal relationship between growth and division was eventually tested extensively in the 1950's by measurements of the growth of individual cells of various kinds as they progressed from one division to the next. This work clearly demonstrated that growth of a cell to some critical size is not the trigger of the cell division. Still, the study of growth through the cycle, particularly in relation to the progression of G_1–S–G_2–D, remains an important part of the analysis of how the cell cycle works. One study of cell growth of mouse L cells in culture (253) has indicated that cell growth may have a causal role in the initiation of DNA replication (see Chapter 4).

It is not a simple task to determine accurately the course of growth over the cell cycle. From a methodological standpoint there are three general approaches. (a) The most direct method is to measure the size of an individual cell as it progresses from one division to the next. This requires a method for accurate determination of the size of single, living cells, and, except for a few cell types, this is technically unfeasible. (b) Theoretically, it should be possible to obtain an accurate growth curve for the cell cycle by making measurements on a large population of cells in which all the cells are progressing synchronously through the cycle. In practice, none of the available methods for obtaining synchrony in cell populations is good enough for obtaining highly accurate growth curves. Limitations of synchrony systems are discussed in the next chapter. (c) Finally, an estimate of the course of cell growth can also be obtained by measuring the distribution of cell sizes in a perfectly asynchronous population of cells. This indirect and somewhat complicated method is the least reliable because of the inadequacies in the instrumentation used to measure cell sizes in large populations and because of variables that cannot be accurately assessed. For example, cell size is not a highly precise indicator of the position of a cell in the cycle. That is, cells at the same position in the cycle show some variation in size. An illustration of this is given in Fig. 24, which shows the variation in size for a homogeneous population of cells in mitosis.

Mazia (307), Kubitschek (272), and Mitchison (318) have assembled comprehensive and careful reviews of the many studies made of cell growth. The principal conclusions that have been derived from such studies are illustrated here by measurements on four cell types: fission yeast, *Amoeba proteus,* *Tetrahymena,* and a mouse fibroblast in culture.

FISSION YEAST (*Schizosaccharomyces pombe*)

Mitchison and his colleagues (318) have studied extensively and carefully the growth of the fission yeast. This cell is convenient for volume measurements because it is cylinder that maintains a constant diameter of about 3 μm and grows by increasing in length (about 8 μm at the start of the cell cycle) (Fig. 3). Hence, it is only necessary to measure the increasing length of the cell between divisions to determine the volume growth curve. The constancy of cell diameter is a valuable advantage since relatively small errors in measuring a changing diameter would result in large errors in determination of volume. Growth in volume of this cell follows a curve with a slightly upward swing (Fig. 4), showing that the rate of volume growth is not constant, but increases as the cycle progresses. Near the end of the cycle volume, growth slows and then ceases altogether. During this period of constant volume the cell plate forms, dividing the cell in two, and the two daughter cells separate.

The course of growth of fission yeast in terms of dry mass, measured by interference microscopy, is significantly different from the course of volume increase (Fig. 4). While growth in volume follows an approximately exponential course, growth in dry mass follows a linear course, i.e., proceeds at a fixed rate. In addition, growth in dry mass does not slow down during cell division. To

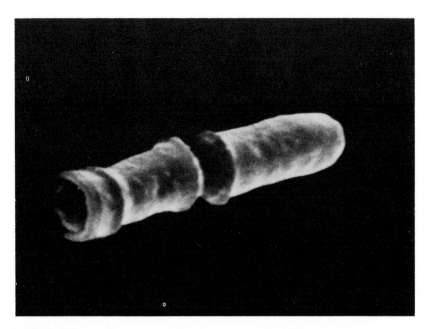

Fig. 3. Scanning electron micrograph of the fission yeast *Schizosaccharomyces pombe* that has just divided. The fission yeast is a cylindrical cell that grows by elongation and maintaining a constant diameter. Photomicrograph by B.F. Johnson, L.C. Sowden, and R.H. Whitehead.

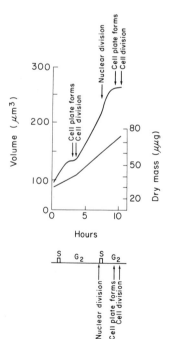

Fig. 4. Volume and dry mass growth of single cells of the fission yeast *Schizosaccharomyces pombe* over the cell cycle. Dry mass was determined by interference microscopy. The experiment was begun with a single cell that subsequently divided, and measurements for the two daughter cells were added together to give a division to division curve. Redrawn from Mitchison (318). The sections of the cell cycle are indicated below [from Bostock (58)] . Nuclear division is immediately followed by a very short S period (10 minutes), and hence there is no G_1 period. Since cell division lags considerably behind nuclear division, the nucleus is already out of S and into the G_2 period by the time cell division is completed.

account for the differences between the way volume and dry mass increase, it must be assumed that water intake occurs at an accelerating rate through most of the cycle and then decreases to zero during cell division.

In pulse labeling experiments the rates of incorporation of radioactive precursors into total protein, total carbohydrates, and total RNA increase with increasing cell size during the cycle in fission yeast (323, 324). If it is assumed that rates of degradation of these macromolecules are constant, the results mean that protein, carbohydrates, and RNA accumulate at increasing rates through the cycle. The apparent inconsistency of these results with the previously found linear increase in total mass was explained by measurement of the size of the precursor pools (322). During the first half of the cycle the total pool size increases faster than does total cell mass, and in the last half of the cycle the pool size decreases relative to total mass. This fluctuation in pool size is sufficient to account for a linear increase in mass (which includes the pool material) in the face of accelerating rates of accumulation of protein, carbohydrate, and RNA.

All these careful measurements on fission yeast, showing that changes in volume, mass, rates of synthesis of macromolecules, and total pool size each follow different courses, illustrate that cell growth does not follow any simple or easily perceived rules.

Amoeba proteus

The growth in mass of individual amoebae from division to division can be followed with the Cartesian diver balance, an instrument devised by Zeuthen (557, 559) for accurate measurement of mass at an ultramicro level. The diver balance consists of a hollow glass sphere whose enclosed space opens to the outside through a long capillary tail (Fig. 5). A plastic cup is mounted on the glass sphere to hold the object (cell) to be weighed. The diver is submerged in a physiological salt solution in a flotation chamber in which it is given buoyancy by an air bubble in the glass sphere. The flotation chamber is connected to a manometer. Decreasing or increasing the pressure in the flotation chamber with the manometer causes the air bubble in the diver balance to expand or contract. By adjusting the pressure carefully the diver can be balanced at a fixed suspended position in the flotation chamber. The diver is first balanced while empty and then balanced after it has been loaded with an object, in this case, an amoeba (Fig. 6). The amount by which the pressure in the system must be reduced in order to expand the air bubble in the diver so that it can support the added weight in the form of an amoeba is an accurate measure of the weight of the amoeba. The instrument actually measures the "reduced weight" of an object, that is, its weight when submerged in a physiological salt solution. The

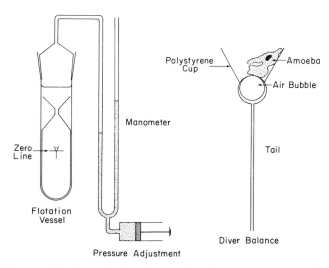

Fig. 5. Diagram of the Cartesian diver balance. Details of the structure of the balance are shown to the right. The balance is shown suspended at the zero line in the flotation chamber to the left.

instrument can be calibrated in terms of mass units by balancing the diver when loaded with an object of a known weight. Diver balances can be constructed, under a dissecting microscope, that are accurate to ±0.01 ng (297).

For the amoeba the rate of cell growth, measured as reduced weight, decreases as interphase progresses, and all growth stops several hours before cell division (Fig. 7) (398). Like fission yeast the amoeba begins DNA synthesis immediately after nuclear division, and therefore a G_1 period is absent. The S period occupies the first 5 or 6 hours of interphase. The nucleus increases rapidly in volume during S, increases slowly in volume through the G_2 period, and swells rapidly during prophase (Fig. 7).

Growth in mass has also been followed in abnormally small and abnormally large daugher amoebae (398). Such amoebae are obtained by subjecting dividing cells to a strong light. Amoebae are negatively phototropic, and inducement of ameboid movement away from a source of light during a late stage of cytokinesis causes unequal distribution of cytoplasm between the two daughter cells. In the subsequent cell cycle the abnormally small daughter cell grows more rapidly than a normal-sized amoeba, and the large daughter cell grows more slowly (Fig. 8). As a result, both daughter cells attain the same weight as the average normal cell by the time of the next cell division. The larger daughter usually has a generation time slightly shorter than normal while the smaller daughter has a longer than normal generation time (see Fig. 19).

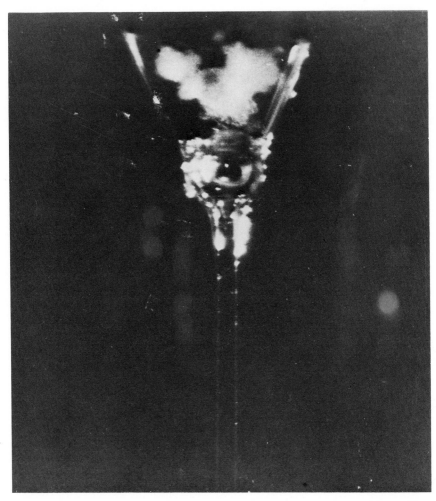

Fig. 6. A Cartesian diver balance loaded with an amoeba. Only a portion of the capillary tail is shown.

A more abnormal growth situation can be created by inhibition of cytokinesis by immersing a dividing amoeba in a 1% albumin solution. Mitosis proceeds normally, and the result is a binucleated cell. Removing one of the daughter nuclei by microsurgery creates a division-sized cell that contains a single nucleus that is just starting the cycle. If such a cell is prevented from growing by deprivation of nutrients, it divides anyway and with a generation time somewhat shorter than a normal growing cell. Thus, a cell that normally grows through most of interphase can transit the cycle without any growth at all, if it is large

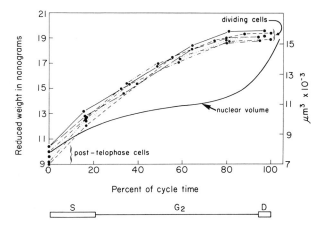

Fig. 7. Growth curves (closed circles) in reduced weight in nanograms, determined with a Cartesian diver balance for six individual amoebae through the cell cycle, and the growth of the nucleus in volume. The sections of the cell cycle are shown below the figure. From Prescott (398).

enough at the beginning of the cycle. The generation time in this case is not determined by a need for growth but instead reflects the time necessary to complete some program of events that includes DNA replication but those other steps are still largely unidentified.

The relation between growth and division in an amoeba has also been examined by stopping the growth of a normal cell part way through the cycle by withdrawal of food (Fig. 9). Such an amoeba will divide anyway, although the generation time is considerably lengthened, and two abnormally small daughter

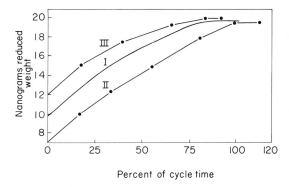

Fig. 8. Growth curves in reduced weight in nanograms for an abnormally large (III) and an abnormally small (II) daughter cell. The growth curve labeled (I) is for a normal-sized daughter amoeba. From Prescott (398).

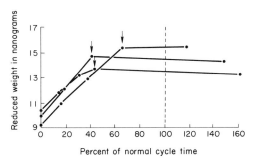

Fig. 9. Three growth curves for amoebae whose growth was stopped by deprivation of food at times indicated by arrows. The cells divided eventually in spite of the limitation on growth but had extended generation times. The ends of the growth curves mark the times of divisions. From Prescott (398).

cells are produced (398). The more severely growth is limited in such experiments, the longer the generation time. Apparently, the necessary events leading to division can be accomplished without the normal amount of growth, but it takes longer. In this respect cell mass is important for the rate at which a cell progresses toward cell division, but attainment of a particular cell mass is obviously not a requirement for initiation of the division process.

These various experiments on amoebae lead to three conclusions. (1) The rate of cell growth is related to cell size; the larger the cell the slower it grows. (2) Cells normally grow to a particular size before division. (3) The initiation of division does not *require* growth of the cell to some particular size. Thus, although growth and division are obviously interrelated, the relationship is rather a loose one.

Tetrahymena pyriformis

Cell cycle growth in *Tetrahymena* has been followed by measurement of the rate of respiration of single cells in a Cartesian diver respirometer (558). The rate of respiration is assumed to reflect the amount of respiratory machinery and, therefore, to be a measure of growth. The increase in respiration rate from division to division is linear (Fig. 10), but during division the respiration rate remains constant or declines slightly, suggesting a cessation of growth during division.

The volume growth of individual *Tetrahymena* has been followed by measuring the increase in area of cells flattened to a constant thickness with a coverglass. Cell volume increases at a constant rate through interphase and then the rate accelerates during cell division (Fig. 11) (95). Thus respiration rate and volume both show linear growth curves through interphase, but indicate very different growth behavior during cell division.

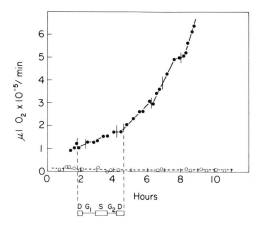

Fig. 10. The increase in respiration rate during the cell cycle in *Tetrahymena* growing in a Cartesian diver respirometer. The experiment was begun with a single cell and ended with 16 cells. The pairs of vertical bars mark the intervals during which cell division occurred. Increasing separation of the two bars of a pair reflects decreasing synchrony in the clone. Redrawn from Zeuthen (558). The sections of the cell cycle, defined by activities of the macronucleus, are indicated below the graph.

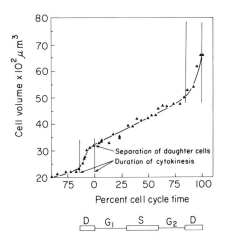

Fig. 11. Volume increases in *Tetrahymena* during the cell cycle. The pairs of vertical bars indicate the period of cell division. Redrawn from Cameron and Prescott (95).

MOUSE FIBROBLAST

Division to division growth curves have been obtained by measuring the dry mouse fibroblasts in an asynchronous culture by interference microscopy (252) in which the postdivision age of individually measured fibroblasts was known from a photographic record of the culture. From this information a growth curve could be constructed (Fig. 12). Although it is clear that cell growth is continuous during the cycle, the growth curve is not sufficiently precise to reveal whether growth occurs at a constant rate or a changing (increasing) rate, although the data tend to suggest an increasing growth rate as the cycle proceeds.

The rate of incorporation of ^{14}C-leucine into fibroblasts of known post-division ages increases steadily through the cycle (Fig. 13) (554). This increase presumably reflects an accelerating growth rate since proteins account for at least 80% of fibroblast dry mass. This interpretation is a little uncertain because the leucine pool size might change over the cycle, resulting in changes in the rate of ^{14}C-leucine incorporation independent of the rate of protein synthesis. In

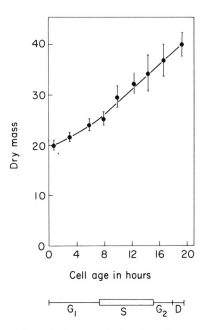

Fig. 12. The course of dry mass increase during the cell cycle constructed from measurement of mass interference microscopy of a large number of mouse fibroblasts of known postdivision ages. Redrawn from Killander and Zetterberg (252).

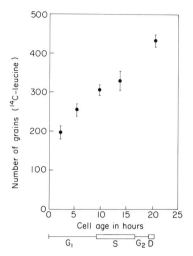

Fig. 13. Rate of protein synthesis measured in mouse fibroblasts of known postdivision age by short incubations with ^{14}C-leucine followed by quantitative autoradiography. Redrawn from Zetterberg and Killander (554). The sections of the cell cycle are shown below the graph.

addition, changes in the rate of protein degradation over the cycle might invalidate the use of the rate of ^{14}C-leucine incorporation as a measure of protein increase.

CONCLUSIONS

Studies of cell cycle growth have not revealed any simple universal rules. Some cells grow at an accelerating rate through the cycle (e.g., volume of fission yeast, protein synthesis in mouse fibroblasts). Other cells grow at a constant rate (macromolecular synthesis in fission yeast; volume and respiration rates in *Tetrahymena*). Finally, at least one cell (*Amoeba proteus*) grows at a decreasing rate through its cycle. We do not yet understand what governs the growth rate of any cell, and, indeed, the pattern of growth may be different when different properties are measured (e.g., volume vs mass). It is obvious that growth is causally interrelated with cell division, and cells left to grow undisturbed will, on the average, double in size prior to division. However, this relationship may be easily altered experimentally by limiting growth. For example, restriction of growth to less than a doubling in size (amoeba experiments) does not prevent division, showing that growth and division are not tightly coupled. Of course, if the restriction of cell growth is severe enough, cell division always ceases.

Finally, patterns of growth show no particular relationship to the progression of a cell through G_1, S, and G_2. For example, in general there is no change in growth rate when cells enter DNA replication. For cells whose nuclei divide mitotically (which includes almost every eukaryote), the growth rate drops dramatically during mitosis (see Chapter 8 for a discussion of RNA and protein synthesis during mitosis).

3

Cell Synchrony

Measurements of properties and activities in single cells are difficult to do, even when they are technically feasible. Indeed, many of the types of measurements needed to define the components of the cell cycle are essentially impossible with single cells. To circumvent these difficulties a variety of methods has been developed to obtain large populations of cells synchronized with respect to the cell cycle. Because these methods vary in effectiveness, it is important to consider their limitations in the interpretation of cell cycle measurements.

There are two general types of synchrony systems; those in which synchrony occurs naturally and those in which synchrony is experimentally derived. Experimentally derived systems consist, in turn, of two types, called selection synchrony and induced synchrony (243). Selection synchrony is the physical separation of cells that are in the same stage of the cell cycle from an asynchronous population. Induced synchrony is usually performed by blocking the progress of the cell cycle so that all cells are brought to the same cycle position. Each of the three synchrony systems has its particular advantages and each has limitations in its usefulness.

NATURAL SYNCHRONY

Sea Urchin Eggs

One of the best systems of natural synchrony extensively used to study cell reproduction is the cleavage of sea urchin embryos. Gram quantities of sea urchin eggs can be fertilized simultaneously, and the synchrony of the first three cleavages is excellent. Mazia and Dan (306) exploited the synchrony of the first cleavages of sea urchin embryos to pioneer the isolation of the mitotic apparatus in bulk for compositional and other analyses. The sea urchin system has been particularly useful to study the shifting of DNA polymerase between nucleus and cytoplasm in relation to the cell cycle (Chapter 5). For some kinds of analyses, dividing sea urchin eggs may be a poor choice since growth of the cells consists of considerably less than a doubling of structural and functional components between divisions. Alternatively, because of the small amount of cell growth, these cells might be particularly advantageous for study of certain cell cycle events, for example, control of initiation of DNA replication.

Physarum polycephalum

Another organism with exceptionally good, natural synchrony extensively used for cell cycle work is the slime mold *Physarum polycephalum. Physarum* is a plasmodium, that is, it consists of a single cytoplasmic compartment containing many thousands of nuclei. All the nuclei traverse the cell cycle in virtually perfect synchrony. Cytokinesis, however, does not occur, and the plasmodium continues to increase in mass with each reproduction of nuclei. A single plasmodium maintained aseptically in defined nutrient medium quickly grows into a sheet with a diameter of many centimeters. A single plasmodium will yield milligram quantities of protein and RNA and several hundred micrograms of DNA (420). There is no G_1 period for the nuclei of *Physarum,* and DNA replication begins as telophase is completed, the S period lasts about 3 hours, the

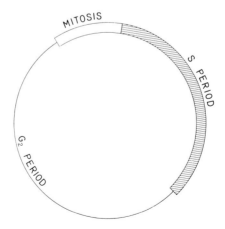

Fig. 14. Cell cycle of *Physarum polycephalum.* DNA synthesis begins at the end of mitosis, and hence, a G_1 period is not present.

G_2 period lasts about 5 hours, and mitosis about 1 hour (generation time = 9 hours) under good growth conditions (Fig. 14). The cell cycle is unusual in that synthesis of nucleolus-associated DNA continues throughout the G_2 period. A variety of experiments on *Physarum*, discussed in subsequent chapters, attests to the unusual usefulness of this organism for many kinds of cell cycle analysis.

EXPERIMENTALLY DERIVED SYNCHRONY

Selection Synchrony

These methods, which have been classified and reviewed by James (243) and Mitchison (318), consist primarily of separation of cells by size (by velocity sedimentation) or by selective removal of dividing cells from cultures.

Velocity Sedimentation. This method exploits the fact that a given size class of cells will all be in *approximately* the same position in the cell cycle. The separation of yeast cells by size (Fig. 15) (321) has been useful particularly for the measurement of enzyme activities in relation to the cell cycle [reviewed by Mitchison (318)] (see Chapter 13).

The velocity sedimentation method has also been applied successfully to mammalian cells (432, 445) (Fig. 16). The degree of synchrony obtained with these methods is good enough to be useful for many kinds of experiments. This method must necessarily remain imperfect because of some variability in the

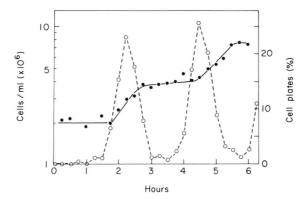

Fig. 15. Growth of a synchronous population of fission yeast obtained by separation of smaller cells from an asynchronous population by velocity sedimentation in a sucrose gradient. Closed circles are cell numbers and open circles show the cell plate index, which is the percentage of cells engaged in formation of a division plate through the middle of the cell. Redrawn from Mitchison (318).

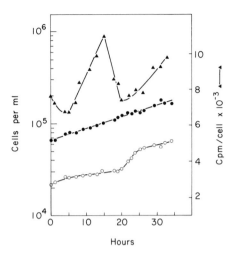

Fig. 16. Synchronous culture of mouse LS cells obtained by separation of small cells from an asynchronous population by velocity sedimentation in a sucrose gradient. The circles show the increase in cell number in an unsynchronized (closed circles) and a synchronized (open circles) culture. The triangles show incorporation of ^{3}H-thymidine into DNA during 30-minute pulses in the synchronized culture. From experiments of McClelland and Shall described in Shall (445).

relationship between cell size and cell cycle position. Under the best of conditions cells of the same size tend to be scattered within a short segment of the cycle rather than precisely aligned at a single point.

Individual Cell Selection. In the earliest methods of synchronization by selection, dividing cells were collected one by one with a fine-tipped braking pipette under a dissecting microscope [for example, Prescott (395)]. The method is limited to a few kinds of cells large enough to work with under a dissecting microscope, for example, *Tetrahymena* (see below), *Amoeba proteus, Euplotes,* and *Paramecium.* The yield of synchronous cells is limited to a few hundred to a thousand because of the labor of single cell selection. The method has been used to measure incorporation of ^{3}H-labeled amino acids into total nuclear proteins and into histones during the cell cycle of the ciliate *Euplotes* (see Chapter 11 and Fig. 58).

Tetrahymena pyriformis, strain HSM, synchronized by individual selection, provided an early demonstration of the rapid loss of synchrony (395) that is encountered following the synchronization of any cell type. The experiment in Fig. 17 was begun with 25 dividing *Tetrahymena* that yielded 50 daughter cells that were essentially in perfect synchrony. When the next cell division was reached, the synchrony of these 50 cells had already deteriorated

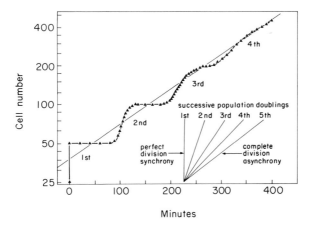

Fig. 17. Multiplication pattern over four cell cycles for a population of 25 *Tetrahymena* synchronized by individual selection of dividing cells. The slopes at the lower right show the loss of synchrony with each successive cell cycle. Redrawn from Prescott (395).

significantly. With the third cell division (increases from 100 to 200 cells) the synchrony was poor, and by the fourth division (200 to 400 cells) almost all trace of the initially perfect synchrony was gone.

The loss of synchrony in the *Tetrahymena* experiment is due to variation in generation time that is always present even though the cells are genetically identical and are growing in the same culture. The distribution of generation times for a clone of *Tetrahymena* proliferating with an average generation time of 111 minutes is shown in Fig. 18. The individual generation times vary from

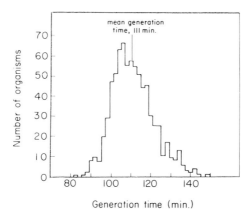

Fig. 18. Distribution of generation times for a clone of *Tetrahymena pyriformis,* strain HSM, all growing under identical culture conditions. Redrawn from Prescott (395).

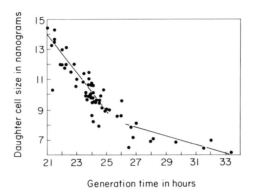

Fig. 19. Relationship between daughter cell size (measured with a Cartesian diver balance) and generation time in *Amoeba proteus*. Redrawn from Prescott (397).

82 to 149 minutes with a slight skew toward the longer generation times. Similar, broad distributions in generation times have been found for bacteria, yeast, algae, and mammalian cells [see Kubitschek (271)]. The wide variation in generation times reflects nongenetic inequalities among the individual cells in a population. A likely basis for such inequalities is the failure of cytokinesis to produce two daughter cells of exactly the same size. This explanation is supported by the observation of an inverse relationship between daughter cell size and generation time in *Amoeba proteus* (Fig. 19).

Mitotic Selection. Currently, one of the most widely used synchrony methods is mitotic selection (373). The method can be applied to most kinds of animal cells that grow as monolayers in culture. During mitosis most kinds of animal cells become roughly spherical and are only loosely attached to the surface of the culture vessel (Fig. 20). Since interphase cells are generally flat and firmly attached to the vessel surface, agitation of the medium results in selective detachment of the dividing cells. By this method a population consisting of up to 99% mitotic cells can be obtained by careful manipulation (Fig. 21).

One disadvantage of the mitotic selection method is the relatively low cell yield. Since even for cells with relatively short generation times only about 4% of the cells in an asynchronous culture are in various stages of mitosis at any moment, the maximum yield from a single large monolayer culture of 20×10^6 cells is theoretically about 8×10^5 cells. In practice the yield of mitotic cells by selection is much less than the theoretical amount. However, the same culture can be used repeatedly every 20 minutes to obtain successive groups of synchronized cells. Alternatively, the mitotic cells can be chilled to $0°C$ as they are collected to hold them in mitosis. The successive yields of mitotic cells can then be pooled and returned to $37°C$ to reinitiate synchronous traverse of the cells

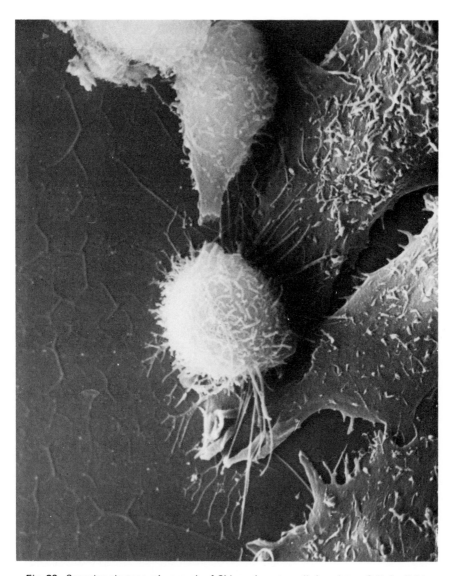

Fig. 20. Scanning electron micrograph of Chinese hamster cells in culture. Cells in division become spherical and are tenuously attached to the surface. Therefore, dividing cells such as the one shown can be preferentially dislodged by agitation of the medium. From Porter *et al.* (384).

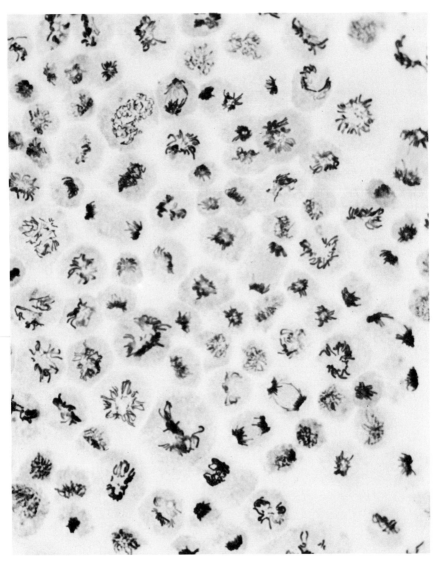

Fig. 21. A sample of hamster cells obtained by mitotic selection. This population consists of late prophase, metaphase, anaphase, and early telophase cells. Late prophase and early telophase cells are only about 20 minutes out of phase, and hence the population is highly synchronous at this point.

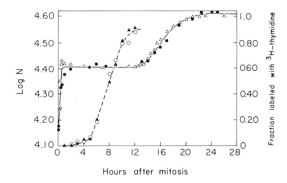

Hours after mitosis

Fig. 22. Comparison of the synchrony of mitotic cells (Chinese hamster ovary line) collected by selection and immediately stored at 0°C for a few hours (circles) with mitotic cells allowed to proceed through the cycle without an interval of cold storage (triangles). Cell numbers are indicated by closed symbols, and entry into DNA synthesis is shown by the open symbols. Both populations proceed through the cycle at the same rate and enter the next division over the same time interval. Redrawn from Enger and Tobey (130).

through the cycle. At least for some cells, e.g., the Chinese hamster ovary cell line, storage of mitotic cells at 0°C for a few hours does not alter the subsequent rate of cell cycle traverse when the cells are returned to 37°C (Fig. 22) (130).

Another way of increasing the yield of mitotic cells is to induce crude synchrony in the population at an earlier point in the cell cycle and then to perform the mitotic selection when the cells subsequently arrive at mitosis. For example, this can be done by treating cells with a high concentration of thymidine, which synchronizes cells in the early part of the S period (see page 33).

Unfortunately, the variation in generation time described above for *Tetrahymena* also occurs in animal cells (Fig. 23). Because of this variation,

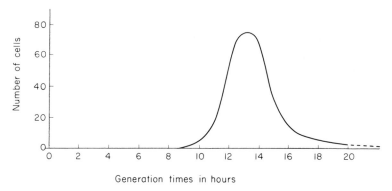

Generation times in hours

Fig. 23. Distribution of generation times of 279 cells of a clone of Chinese hamster ovary cells determined by time-lapse photography.

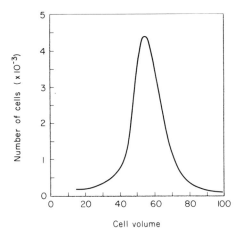

Fig. 24. Distribution of cell volumes for mitotic cells of a Chinese hamster cell line determined by an electronic particle sizer. Redrawn from Anderson *et al.* (14).

synchrony of mitotically selected cells deteriorates markedly during the subsequent cell cycle. The extent of this deterioration is shown in Fig. 22. The initially, tightly synchronized mitotic cells divide over a spread of about 10 hours at the next division. Most of the loss in synchrony is attributable to the high variability of the rate at which the cells traverse the G_1 period (see Chapter 4). Variableness in the length of G_1 may, in turn, be due to variation in daughter cell size at the beginning of the cycle. The size (mass) of daughter cells of a line of mouse fibroblasts has been shown to vary over a twofold range. This size variation probably stems, at least in part, from unequal cell division. In addition, considerable size variation is present in mitotic cells themselves (14). Figure 24 shows the distribution of cell volumes of mitotic cells, obtained by selection, for a line of hamster cells. The size variation covers a twofold range. This broad distribution in the size of dividing cells also bears on a point raised earlier about cell growth and the initiation of cell division, affirming the conclusion that cell division is not initiated by growth of a cell to a particular size.

In mitotically selected cells the manipulation of the cells is not the cause of the loss of synchrony. This is known because the distribution of generation times shown in Fig. 23 was determined by time-lapse photography of a cell culture growing in an undisturbed culture, and this distribution of generation times is sufficient to account for the rate of synchrony loss.

It is evident that variation in generation times does not reflect heritable differences among the individual cells because the average generation time for a cell line usually remains constant from one subcultivation to the next. If the differences were heritable, the average generation time would soon decrease to the generation time of the fastest cell.

Retroactive Synchrony. With the conventional mitotic selection method, the synchrony is at its maximum early in the cycle. Thus, measurements of cell cycle phenomena in mitotically selected cells become progressively less precise as the cycle progresses. The deterioration in synchrony is so great that the method is essentially useless for the study of the G_2 period. An alternative form of the mitotic selection method called *retroactive synchrony* circumvents this difficulty for certain kinds of experiments. Retroactive synchrony is similar, in principle, to the method originally developed by Howard and Pelc (237) to define the four sections of the cell cycle. Howard and Pelc developed what is now usually called the *labeled mitotic index method* for determining the average length of the four sections of the cycle for cells in tissues or in culture. In the labeled mitotic index method an asynchronously proliferating cell population is given a short pulse of ^{3}H-thymidine. Samples of cells are prepared for microscopic examination at regular intervals after the pulse, and the percentage of mitotic cells that is radioactive in the successive samples is determined by autoradiography. For the first several hours after the pulse of ^{3}H-thymidine all cells arriving at mitosis are not radioactive (Fig. 25). These are cells that were in the G_2 period when the ^{3}H-thymidine was available. As cells which were in the latter part of the S period begin to enter mitosis, the percentage of mitotic cells that is labeled begins to rise. The length of the G_2 period varies somewhat from cell to cell, and the average length of G_2 is taken as the time interval between the ^{3}H-thymidine pulse and 50% labeled mitotic cells (Fig. 25). As cells labeled during slightly earlier stages of the S period arrive in mitosis, the percentage of labeled mitotic cells rises to 100%. Eventually cells which were in G_1 during the pulse of ^{3}H-thymidine begin to arrive in mitosis and the labeled mitotic index begins to fall. The average length of the S period is taken as the time interval between the ascending and descending slopes at the point of 50% labeled mitotic cells.

The first labeled cells to reach mitosis eventually pass through a full cell cycle and arrive once more in mitosis. The labeled mitotic index therefore rises again. An estimate of the average generation time is given by the interval between the 50% points on the two ascending slopes (Fig. 25), although it is prudent to obtain a measure of generation time by an independent method when possible, for example, by measuring the doubling time for the cell population. The time taken for cell division, typically about 1 hour for mammalian cells, can usually be estimated by microscopic observation. Since the average generation time and average lengths of G_2, S, and D are known, the average length of G_1 can be calculated by subtraction.

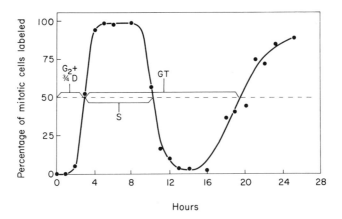

Fig. 25. Determination of the average length of the generation time, G_2, S, and G_1 periods, by the labeled mitosis method. In this plot most of the division period is included with the G_2 period, and the indicated value for G_2 should be corrected by subtraction of about 45 minutes.

To obtain precise results with the labeled mitotic index method, labeling with ^{3}H-thymidine must be held to a minimum (for example, 0.1 μCi/ml of medium for 10 minutes). Even moderate incorporation of ^{3}H-thymidine causes considerable mitotic delay and other cell cycle alterations due to radiation damage and decreases the percentage of cells that reach the second division. Light labeling of DNA has the disadvantage that autoradiographic exposure times must be long. For the experiment described in Fig. 25 the exposure time was 3 months.

In retroactive synchronization the same procedure is used as in the labeled mitotic index method, except that groups of cells are collected by mitotic selection as they arrive at mitosis. With this method it is possible to collect large groups of cells that were pulse labeled at particular times in interphase, for example, with ^{3}H-thymidine, ^{3}H-uridine, and ^{3}H-amino acids, to determine when particular macromolecules were synthesized in the cycle. The first groups of labeled cells to arrive in mitosis represent the maximum "synchrony." Cells from earlier and earlier parts of interphase that arrive in mitosis are progressively less well "synchronized" populations because of the variation in the lengths of G_2, S, and G_1 periods for individual cells. Because the variations in S and G_2 are relatively small, however, the synchrony remains good until the G_1 cells begin to arrive at mitosis. By combining the usual mitotic selection method with the method of retroactive synchrony, it becomes possible to do analyses on the cell cycle from both ends, and thereby reduce to some degree the problem of synchrony deterioration through the cycle.

Induced Synchrony

Induced synchrony involves bringing all or most of the cells in an asynchronous population to a single point in the cycle by manipulation of culture conditions.

Repetitive Heat Shocks. Synchronization of eukaryotic cells was pioneered by Zeuthen and his colleagues, who synchronized cell division in *Tetrahymena* by a series of closely spaced heat shocks (431). Unfortunately, although cell divisions are synchronized, DNA replication is not synchronized, and the system has not been very useful for producing synchronous cell populations for analysis of the cell cycle. However, the mechanism of how heat shocks induce division synchrony, which has been studied extensively by Zeuthen, has in itself revealed significant information about properties of the cell cycle (see 555). Zeuthen (555) has recently achieved a good degree of synchrony of DNA synthesis in *Tetrahymena* by combining the heat shock treatment with reversible inhibition of DNA synthesis with methotrexate.

Inhibition of DNA Synthesis. Several inhibitors of DNA replication can be used to induce the synchronization of the cell cycles of animal cells. Fluorodeoxyuridine (FUdR) inhibits thymidylate synthetase activity and thereby prevents the synthesis of thymidine monophosphate (TMP) from deoxyuridine monophosphate (Fig. 26). Since the pool of thymidine nucleotides in animal cells is sufficient for no more than a few minutes of DNA replication [see, for example (136)], addition of FUdR results in rapid inhibition of DNA synthesis. Cells already in the S period are usually lethally injured by FUdR, but cells in G_2, D, and G_1 apparently proceed normally to the G_1–S border, where they are accumulated. These cells can be released from the FUdR inhibition by addition of thymidine to the medium. Thus, FUdR selectively kills cells in S and synchronizes the non-S cells at the G_1–S border. The method is useful, but always carries the risk that the FUdR treatment has altered the behavior of non-S cells in unperceived ways.

In a similar manner animal cells can be synchronized by inhibiting DNA synthesis with hydroxyurea or cytosine arabinoside (Fig. 26). Hydroxyurea inhibits nucleoside diphosphate reductase and therefore prevents production of deoxynucleoside diphosphates from nucleoside diphosphates. Cytosine arabinoside appears to inhibit DNA polymerase. Both inhibitors kill S-period cells when applied for a few hours; cytosine arabinoside is the more toxic of the two. The inhibitory effect of hydroxyurea on cells accumulated at the G_1–S border is quickly reversed by washing the cells. Inhibition by cytosine arabinoside is reversed much more slowly. There is some doubt that these agents achieve an

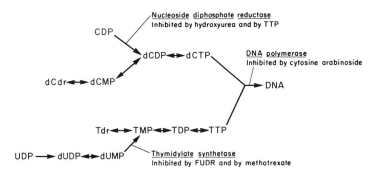

Fig. 26. Block of DNA synthesis by inhibition of the formation of dCTP or of TTP or by inhibition of DNA polymerase.

absolute block at the G_1–S border. Cells finishing G_1 possibly begin DNA synthesis but at a rate that is too low to be readily detected. A main reason for this suspicion is the observation that cells blocked by hydroxyurea or cytosine arabinoside eventually leak far enough into the S period to produce a detectable increase in DNA content.

Double Thymidine Block (49, 120, 160, 254, 484, 543). A high concentration of thymidine added to the medium (typically 2 mM; higher concentrations are injurious to at least some cell types) drastically slows DNA replication (43, 57) in animal cells by inhibiting nucleoside diphosphate reductase (120, 330, 331, 543). The addition of 1 mM thymidine to Chinese hamster ovary cells causes the pools of TTP, dGTP, and dATP to increase about twenty-five-, ten-, and twofold, respectively (43). The dCTP pool is, paradoxically, only slightly decreased. The high intracellular concentration of dTTP is the cause of the slowdown in DNA synthesis through its inhibition of nucleoside diphosphate reductase (Fig. 26). The inhibition of DNA synthesis can be prevented or reversed by deoxycytidine, since deoxycytidine obviates the need for reduction of cytidine diphosphate to deoxycytidine diphosphate.

High thymidine (2 mM) treatment is not lethal for at least several hours for cells in the S period, apparently because the inhibition of DNA replication is not as severe as it is with FUdR, hydroxyurea, or cytosine arabinoside. Presumably, as long as the S period can progress slowly, the cells are protected from permanent damage.

Maximum synchronization is achieved by using two periods of thymidine inhibition separated by a carefully timed release period. The first block is applied for an interval equal to the combined lengths of G_2, D, and G_1. This allows the cells in these sections to proceed to the G_1–S border. Cells in S are severely slowed, and few cells leave the S period during the block. Thus, the first block results in accumulation of the G_2, D, and G_1 cells at the G_1–S border while S period cells remained trapped in S.

The block is reversed by washing the cells free of thymidine. The release, which occurs immediately, is maintained for an interval equal to the length of a normal S period. (If G_2 + D + G_1 is less than the length of S, the double block method cannot be used for that particular cell type, since some cells will reenter the next S period during the release.) The release interval allows all of the cells to complete DNA replication and distribute through G_2, D, and G_1. A second block is applied for an interval sufficient to allow all cells to reach again the G_1–S border. Release of the second block allows the cells to enter DNA synthesis synchronously.

Theoretically, the double thymidine block method should tightly synchronize all cells at the G_1–S border. In fact the method produces only fair synchrony, probably because the block and release times are based on average values for the sections of the cycle, and cells that deviate much from the average will not be efficiently synchronized. Also, since high thymidine does not completely block DNA replication, this method does not achieve an arrest of cells at the G_1–S border as often supposed, but causes them to accumulate in an early part of the S period (43, 57, 313).

Mitotic Selection Plus Inhibition of DNA Synthesis. The synchrony induction methods using FUdR, hydroxyurea, cytosine arabinoside, and high thymidine can be effectively combined with the mitotic selection to obtain highly synchronous cell populations near the G_1–S border. An inhibitor of DNA synthesis is added to cells obtained by mitotic selection, and they are arrested near the G_1–S border. Inhibition with high thymidine or hydroxyurea is preferred because these inhibitors are the least toxic, and the inhibition can be rapidly reversed.

Synchronization in G_1 by Amino Acid Deprivation. A method for synchronization of cultured animal cells has been developed using deprivation of isoleucine or leucine to arrest cells at a point in G_1, followed by reprovision of the amino acid to release the cells (498). Unfortunately, the cells do not enter DNA synthesis any more synchronously than cells synchronized by mitotic selection (Fig. 22). However, like mitotically selected cells, resynchronization at the G_1–S border can be achieved by relieving the amino acid deprivation and at the same time adding hydroxyurea to block DNA synthesis. Release from hydroxyurea then provides a cell population that proceeds through S, G_2, and D with a good degree of synchrony (Fig. 27). With this method an entire culture of cells may be synchronized, overcoming the disadvantage of the limited number of cells yielded by the mitotic selection method.

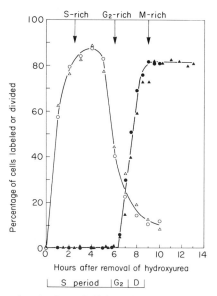

Fig. 27. Two-step synchronization of Chinese hamster cells at the G_1–S border. Cells were first deprived of isoleucine to bring about arrest in G_1. The arrest was reversed by addition of isoleucine and hydroxyurea added at the same time. The graphs show the percentage of cells making DNA (open symbols) after removal from hydroxyurea and the subsequent increase in cell number (closed symbols) for two experiments. Redrawn from Gurley *et al.* (196).

Release of Animal Cells from Density-Dependent Inhibition of Growth.   Normal animal cells in culture become arrested in G_1 (and enter G_0) when the cells proliferate to a confluent monolayer (192, 345). The cells can be released from arrest by subculture into fresh medium or by addition of fresh serum. Some hours after release, depending on the cell type and the length of time the cells have been arrested, a high percentage of the cells will enter DNA synthesis with a fair degree of synchrony (Fig. 28). The method is useful mainly for the study of reversal of the G_0 state and of events leading to DNA synthesis.

Several other methods of induced synchrony have been developed for particular cell types, for example, light–dark cycles to induce and maintain synchrony of algae. Some of these methods are reviewed by Mitchison (318), and an extensive coverage can be found in books edited by Cameron and Padilla (93) and by Zeuthen (556).

These various synchrony systems have greatly extended the possibilities for cell cycle analysis well beyond what can be done with single cells, although certain kinds of single cell experiments, for example, experiments based on nuclear transplantation or cell fusion (Chapter 5), continue to have their own unique value. Although the current methods for experimentally derived syn-

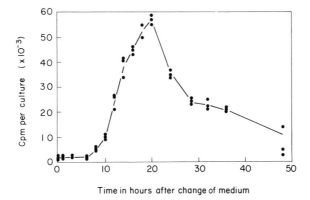

Fig. 28. Rate of incorporation of ³H-thymidine into DNA in mouse 3T6 cells after release from growth arrest by replacement of old medium with fresh medium. Each point was obtained by labeling a cell culture with ³H-thymidine for 30 minutes. Redrawn from Tsuboi and Baserga (511).

chrony are important tools in the contemporary study of the cell cycle, it is also apparent that all these methods are limited in their usefulness. This is partly because they produce imprecisely synchronized cell populations and partly because individual cell variations cause rapid deterioration of synchrony. Unfortunately, no method yields synchrony of sufficient precision to permit fine scale resolution of temporally ordered events in the cell cycle, such as events at the G_1-S border.

4

The G₁ Period

VARIABLENESS OF G_1

Although the G_1 period has not yet been explained by any specific events, some of its important properties have been identified. One of the first clues about the significance of the G_1 period came with the observation, already introduced in Chapter 3, that in a homogeneous population of cultured cells the G_1 period is far more variable in length than S, G_2, or D. This was clearly pointed out in the early studies of Sisken and Kinosita (455) on cultures of human and cat cells and has since been documented numerous times for the cycles of a variety of cultured cells.

The wide variableness in the lengths of the G_1 periods for individual cells gives rise to most of the variableness in generation times within a cell population. An example of this relationship is illustrated in Fig. 29 for a clonal line of Chinese hamster ovary cells. For 279 cells the generation times varied from 10 to 20 hours with an average cycle time of 13.5 hours. The length of S + G_2 + D for 141 cells varied from 9 to 11 hours with an average of 9.6 hours. From the difference between the two sets of values, the G_1 period was estimated to range from about 1 to 9 hours with an average G_1 of 3.9 hours.

Little is known about the possible cause of the variableness in G_1 among individual cells in a homogeneous population. It is not possible to decide, for example, whether this variableness represents a variable rate with which cells move through some part or all of the G_1 period or whether it stems from a transient arrest of variable duration at some particular point in G_1. The idea of a transient arrest of variable duration stems from the observations that a frank but reversible arrest of cells in G_1 (entry into G_0) is the mechanism by which cell

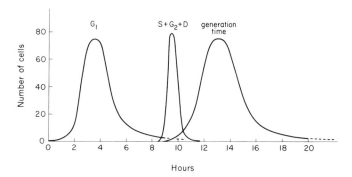

Fig. 29. A comparison of the variableness in lengths of the G_1 period, $S + G_2$ + D, and generation time. The two right-hand curves show the experimental measurements of the distribution of generation times determined by time-lapse photography and the distribution of values for $S + G_2$ + D determined by the labeled mitotic index method for a line of Chinese hamster ovary cells grown in monolayer. The left-hand curve is the estimate of the variableness in G_1 times obtained by subtracting the curve for $S + G_2$ + D from the curve for generation times.

reproduction is regulated. Examples of regulation by G_1 arrest are the specific inhibition of cell reproduction in tissues, density-dependent inhibition of growth in cultured cells, and inhibition of cell reproduction caused by nutrient deprivation of cultured cells in general. The data supporting the concept of a checkpoint in G_1 at which reversible cell arrest occurs are discussed in detail in a later section. On the basis of the evidence of G_1 arrest, it becomes reasonable to speculate that variableness in the length of G_1 in an actively growing culture reflects a tendency for cells to be transiently retained, for variable durations, at that checkpoint in G_1 at which the cycle is interrupted to achieve regulation of cell reproduction.

The variableness of G_1 is readily perceived with synchronous cells obtained by the mitotic selection procedure. In the hours between completion of mitosis and the initiation of the S period, the decay in synchrony is severe. In the experiment in Fig. 30, a cell population of Chinese hamster ovary cells (CHO) consisting of 99% mitotic cells was obtained by mitotic selection. The mitotic cells were maintained at a constant temperature and were allowed to settle in a culture vessel without centrifugation in order to minimize disturbances that might cause G_1 variableness. To obtain a cell population with a mitotic index of 99%, however, requires that the monolayer be subjected to several preshakes to remove any loosely attached interphase cells. This involves replacing the old (conditioned) medium with fresh medium just prior to the final mitotic selection. The medium change could conceivably affect the average length and degree of variableness of the G_1 period in the subsequent cycle. In any case, the time of entrance of the cells into S in Fig. 30, assessed by autoradiographic detection of

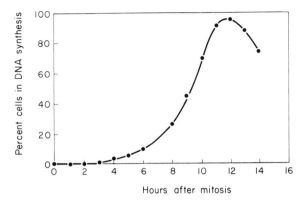

Hours after mitosis

Fig. 30. The curve describes the entry into DNA synthesis of CHO cells obtained by mitotic selection. Because of variableness in the length of the G_1 period for individual cells, the loss of synchrony has become severe by the time cells reach the S period.

[3]H-thymidine incorporation, extended from 4 to 12 hours after mitosis. By 12 hours after mitosis the cells that began DNA replication at 4 hours were leaving the S period. Even when the selection of CHO cells is done with great care, the average G_1 period (8 hours) is longer than the average G_1 for cells in an undisturbed monolayer (3.9 hours), presumably as a result of the change in medium and the disturbance of mechanical shaking of the cells. The range of G_1 variableness, however, is about the same in both situations (4 to 12 hours for mitotic selection vs 1 to 9 hours for undisturbed cells).

The disturbances to the normal behavior of cells mechanically selected at mitosis is particularly clear in the case of a G_1-less line of Chinese hamster cells known as the V79 line (409). V79 cells growing undisturbed in monolayer have a generation time of 8 to 9 hours with no measurable G_1 period. Mitotic cells obtained by mechanical selection, however, have a G_1 period ranging from 1 to 6 hours in the first cycle after the shake-off. In the subsequent cycle the G_1 period is again essentially zero in all cells. A G_1 period (or G_0 state) does appear normally in V79 cells that enter stationary phase of culture growth as a result of depletion of one or another nutrient in the medium. In contrast to the change in the average length of G_1 induced by mitotic selection in the CHO and V79 cell lines, the average length of $S + G_2 + D$ is not measurably affected.

Significantly better synchrony of entry into S can apparently be obtained if mitotically selected cells are seeded at a sufficiently high density (119). The first cells entering S may release a factor(s) into the medium that accelerates the entry into S of cells still in G_1. This is an important observation that needs to be followed up, particularly regarding the identification of the putative stimulatory factor(s) released by the S phase cell.

Obviously, however, mitotic selection of cells, even with the better maintenance of synchrony obtained by planting the cells at a high density, has limited usefulness for the study of the fine details of events close to or subsequent to the G_1–S transition. As mentioned in Chapter 3 mitotically selected cells can be resynchronized near the G_1–S border with an inhibitor of DNA synthesis. Using the currently available inhibitors, however, the cells probably progress slightly into the S period (313, 499). Hence, such cells are still not suitable for analysis of the G_1–S transition.

RELATION OF CELL GROWTH TO THE LENGTH OF G₁

Killander and Zetterberg (252, 253) have concluded from studies on cultured L cells that the variableness in the length of G_1 may be due to variableness in the mass of cells at the beginning of the G_1 period. As already discussed in Chapters 2 and 3 the variableness in the mass of individual cells at the start of G_1 in a homogeneous population is probably due to the variable size of mitotic cells (Fig. 25) and to the usual failure of cytokinesis to divide the cells into two equal-sized daughters. Killander and Zetterberg (253) observed a correlation between the mass of a daughter cell and the percent mass increase during G_1 (Fig. 31); the larger the daughter cell, the less its mass increase during G_1. In addition, the percent mass increase is related to the length of G_1 (Fig. 32); the greater the mass increase in G_1, the longer the duration of G_1. Thus the smaller the cell at birth, the more it grows and the longer its G_1 period. These data fit

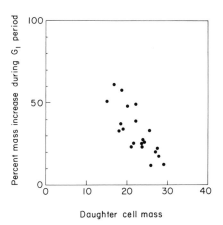

Fig. 31. Mass of individual daughter cells in arbitrary units plotted against their percent mass increase during the G_1 period. Redrawn from Killander and Zetterberg (253).

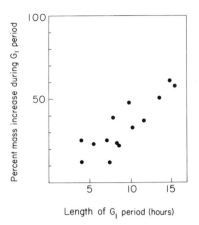

Length of G_1 period (hours)

Fig. 32. Length of the G_1 period for individual cells plotted against their percent mass increase during G_1. Redrawn from Killander and Zetterberg (253).

with the additional observation that the variation in mass among newly divided mouse L cells is significantly greater than is the variation in mass of cells at the beginning of the S period (252).

These studies lead to the conclusion that the initiation of DNA replication is tied to the attainment by a cell of a crucial mass. A very similar hypothesis has been developed for *Escherichia coli* (121), based upon a variety of experiments that indicate that DNA replication is coupled to cell size. It is perhaps more likely that the initiation of DNA synthesis is not governed by the attainment of a given cell mass or even by the attainment of a given, total protein content, but rather by some relatively specific component of growth (for example, ribosome accumulation) that increases in parallel with cell mass.

Fox and Pardee (150) have tested the conclusion of Killander and Zetterberg in another experimental arrangement, using Chinese hamster cells. They obtained a population of mitotic cells by the mitotic selection method and separated the resultant daughter cells into size classes by centrifugation on a Ficoll density gradient. In contrast to the results of Killander and Zetterberg on L cells, only a slight correlation was observed between the sizes of new G_1 cells and the subsequent lengths of the G_1 periods, and the correlation appeared to be too small to account for the high degree of variableness in the duration of individual G_1 periods. The apparent discrepancy may be a result of the techniques used. Killander and Zetterberg used interference microscopy to measure mass and cytophotometry to measure DNA in fixed cells whose ages in the cell cycle were known from time-lapse photography. The only disturbances to the living cells were the photographs taken at 45-minute intervals. The experiments of Fox and Pardee involved mitotic selection and subsequent separation of cells of different sizes on a Ficoll density gradient. Mitotic selection in itself can disturb cell behavior as evidenced by an increase in the average length of the G$_1$ period. It is likely that the separation procedure with a Ficoll gradient, involving changes in medium and other manipulations, even further disrupts the normal behavior of cells. It is quite conceivable, therefore, that the cells in these experiments have been sufficiently disturbed to obliterate a measurable relationship between cell size and the initiation of DNA replication.

The relationship between cell mass and initiation of DNA synthesis has been examined in another way (148). Baby hamster kidney cells (BHK) arrested in G$_1$ (G$_0$) by serum deprivation were stimulated to resume the cell cycle by addition of serum, but were prevented from traversing the S period by the addition of 1 mM hydroxyurea. The cells delayed in S continued to grow, however, to reach abnormally large size as measured by protein content. Large cells with 1.13 ng of protein at mitosis divided into daughters that had the same average G$_1$ period as daughter cells produced from mitotic cells with only 60% as much protein (0.67 ng) (Fig. 33).

Another perspective on cell size and initiation of DNA synthesis has been provided by comparing the cell cycle of haploid and diploid frog cells. If cell size alone were the key to initiation of DNA synthesis, then haploid frog cells, which have half the volume and half the mass of their diploid counterparts, should behave differently than diploid cells. Yet haploid and diploid frog cells have S

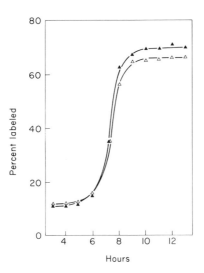

Hours

Fig. 33. G$_1$ length of cells differing in protein content by 60%. The curves show entry into S of cells derived from mitotic cells with 0.67 ng of protein (solid triangle), and cells derived from mitotic cells with 1.13 ng of protein (open triangle). Redrawn from Fournier and Pardee (148).

periods that occupy the same relative positions in the cell cycle and have the same durations (180). Thus, when haploid cells initiate DNA synthesis, they have only about half the size of diploid cells at the beginning of the S period. Obviously, the role of cell mass or cell growth in initiation of DNA replication is not simply achievement of a critical cell size, but involves more complex nuclear–cytoplasmic interactions.

Since a correlation between cell size and DNA replication could be an important clue about the events of G_1 and the control of DNA replication, it is necessary to resolve the differences between the two kinds of results by further experimentation.

As will be discussed later, several experiments have shown unequivocally that the cytoplasm is involved in the initiation of DNA synthesis. It is conceivable that this involvement is dependent upon the amount of cytoplasm, or more precisely, upon the ratio of cytoplasmic size to nuclear size. This possibility has been verified in micrurgical experiments on *Stentor* (153) in which DNA synthesis had ceased because of starvation (presumably arrested in G_1). Increasing the cytoplasmic/nuclear ratio, either by grafting on extra cytoplasm from another starved *Stentor* or by removing part of the polyploid macronucleus, leads to initiation of DNA synthesis. By analogy, these experiments support the idea of Killander and Zetterberg that initiation of DNA synthesis in cultured mammalian cells is triggered by attainment of a particular cell size (cytoplasmic volume).

Other experiments (255) suggest that the initiation of DNA replication may be related to the *rate* of protein synthesis rather than to the net increase in the cell's content of protein (or cellular mass). Thus, under steady-state culture conditions it is reasonable to postulate that the rate of protein synthesis is proportional to the size of the cell. In such a case, the initiation of DNA synthesis might appear to be correlated with the attainment of a paricular mass or protein content, whereas in fact, the crucial factor in the initiation of DNA synthesis might instead be the attainment of a crucial rate of synthesis of total protein or of synthesis of one particular protein. When cells are not in balanced growth or are disrupted in some other way, the rate of protein synthesis may change such that it is no longer correlated with cell size. In such a case the correlation between cell size and the initiation of DNA synthesis may disappear while a postulated relationship between the rate of protein synthesis and the initiation of DNA replication may be present. Thus, the *rate* of protein synthesis might still increase to the crucial level necessary to initiate DNA replication independently of the attainment by the cell of a particular mass or protein content. However, there are no specific clues to suggest how the *rate* of cellular protein synthesis could be directly linked to the control of DNA replication.

Some indirect evidence discussed below supports the idea that the variableness in the lengths of G_1 periods for individual cells, under conditions in which the average G_1 period in the population remains constant, may be based on a variable delay of one or another of the events that are required for transit through a specific part of early G_1. Such delays or hesitations could be caused, for example, by transient deviations within the individual cells from some precise intracellular condition, for example, conditions for synthesis of one or another key protein, that must be fulfilled in order for a particular G_1 event to take place. In addition, environmental changes could impinge on such an arrangement, causing an increase or decrease in the *average delay* at some specific point in G_1 and thereby accounting for the well-known dependence of *average generation time* in a cell population on environmental conditions. A particularly clear example of this is provided by experiments in which CHO cells grown with different lots of sera in the culture media had average generation times of 13.3, 16.5, and 24.1 hours (501). These shifts in generation times were accounted for by changes in the average length of the G_1 period with little or no change in S, G_2, or D. Thus, it is conceivable that the variableness of G_1 length around a constant average and the changes in the average generation time when the environment is changed may both be accounted for by the same G_1 event(s).

This flexibility in the average duration of the G_1 period can be observed in exaggerated form during the transition of a culture from log phase to stationary phase. As the rate of cell proliferation slows, the average duration of the G_1 period increases, usually with relatively much less increase in the lengths of S, G_2, or D [see, for example (203, 255)]. Finally, in true stationary phase, cells remain blocked at some point in G_1 until returned to an environment favorable for cell reproduction. Thus, for example, in experiments on murine lymphoma cells during different parts of the culture growth cycle "the prolongation of population doubling time was mainly due to an extension of the G_1 period, whereas the duration of S, G_2, and mitosis was much less affected" (101). The same is true for lymphocytes of the mouse thymus (139). Similarly, when "stationary phase" was induced by carbohydrate starvation in populations of cells from pea root tips in culture, the initiation of DNA synthesis was increasingly delayed as stationary phase was approached, but once a cell entered DNA synthesis the transit times through S and G_2 were not affected (514).

While a shift in the average length of G_1 is ordinarily the basis for a shift in generation time for cultured cells, occasionally the average lengths of S, G_2, or D are also observed to vary markedly. Lala and Patt (282) found a generation time of 8 hours in 1-day-old Ehrlich ascites tumor cells with S = 6 hours, G_2 + mitosis = 2 hours, and G_1 = 0. In contrast, in 7-day-old ascites tumors the average generation time was 22 hours, S was 18 hours, G_2 + M was 4 hours, and G_1 was still essentially zero. Thus, in this extreme case, an increase in generation time from 8 to 22 hours was due primarily to an increase in the length of the S period with no contribution by a change in G_1 length.

A variety of other times for the sections of the cell cycle of ascites tumor cells has been described (33, 116, 508) in which the generalization about the variableness of G_1 and the constancy of $S + G_2 + D$ does not hold. Some of these exceptions, and possibly all, are due to variable and probably suboptimal nutritional environments provided by the peritoneal cavity of the animal host. For virtually all other kinds of tumor cells (for example, 311) and normal cells growing in culture or in an animal, the generalization about G_1 variableness and the relative constancy of $S + G_2 + D$ is valid.

CONTROL OF CELL REPRODUCTION IN G_1

Reversible Arrest of Cultured Cells in G_1

Another indication of the significance of the G_1 period is provided by the observation that the cessation of cell reproduction in cultures normally occurs by the arrest of the cells in the G_1 state (175, 203, 288, 507). When cell reproduction slows down as a culture enters stationary phase, the G_1 period lengthens until finally the nonreproducing cells remain blocked in the G_1 phase. It is possible to cause log phase cells to become arrested in S or G_2 by the abrupt inhibition of protein or RNA synthesis or by the sudden imposition of some other unfavorable condition in the culture, but the more gradual development of unfavorable conditions, as occurs during the transition to stationary phase in culture, results in G_1 arrest. This suggests that one or more G_1 events essential for advancement in the cell cycle are significantly more sensitive to inhibition by unfavorable or inadequate growth conditions in the cellular environment in stationary phase cultures than are any of the events in S, G_2, or D.

The generalization about G_1 arrest of cultured mammalian cells carries over to the phenomenon of density-dependent inhibition of cell reproduction (192, 345, 506, 548). Whatever the mechanism by which cells mutually restrict their reproductions, the inhibitory signal does not affect S, G_2, or D, but impinges on some essential activity of the G_1 period and prevents the advancement of the cell cycle into DNA replication. The same appears to be true in the interruption of the cell cycle by the inhibitors of cell reproduction present in liver extracts (25). It is possible that the G_1 arrest caused by unfavorable growth conditions, the G_1 arrest in density-dependent inhibition, and the G_1 arrest caused by the inhibitor from liver all act through the same sensitive target mechanism in G_1.

It is well known that in the release of density-dependent inhibited cells with fresh serum many hours are required between the time of addition of serum and the entry of cells into the S period (32, 84, 505, 511). The release of cells from this arrest apparently requires only a brief treatment with fresh serum, and cells are then able to complete the G_1 period in the absence of the original stimulus. From the temporal relationship between release and subsequent initiation of the

S period, it has been argued that cells in density-dependent inhibition are arrested at an early point in G_1. However, this interpretation is complicated by the fact that changes take place in the arrested cell that put it into a divergent G_0 state. Thus, it may be that an arrested cell is in fact in late G_1 but that the cell has entered a G_0 state, and several hours are then required to reinstate it into the cycle.

One indication that arrest of a cell in G_1 diverts it into a G_0 state derives from the observation that the longer a cell is arrested in G_1, the more time it requires to resume proliferation following release of the arrest. For example, as a culture of *Tetrahymena* enters stationary phase, the cells become arrested in macronuclear G_1 (332). The longer the cells are held in stationary phase, the longer the lag period before resumption of cell proliferation when the cells are transferred to fresh medium (396), and hence the deeper they have moved into a G_0 state.

Normal human fibroblasts (WI-38) behave basically the same way (24). Cultures were initiated with a cell inoculum so that confluency was reached in 5 days. Like other density-dependent cells, WI-38 cells arrest in G_1 at confluency (192). When stimulated with fresh medium, these newly arrested cells began DNA synthesis in 8 hours. Cells arrested for 4 days required 14 hours to begin DNA synthesis when released by fresh medium. Cells arrested for 13 days required 20 hours to reach DNA synthesis.

Biochemical proof that cells sink deeper and deeper into a G_0 state when arrested has been provided by measuring the ability of chromatin prepared from arrested cells to support RNA synthesis when provided with RNA polymerase from *E. coli* (24). Chromatin from WI-38 cells that have just entered confluency does not increase in template activity when the cells are placed in fresh medium. Chromatin from cells held in arrest for sometime has less template activity for RNA synthesis than growing cells, and when the cells are released from arrest, there is a lag in the recovery of template activity. It appears that the longer the duration of arrest, the longer the lag in recovery of template activity [see experiments and discussion in Augenlicht and Baserga (24)].

Thus, the fact that cells arrested in G_1 enter a G_0 state and sink deeper into G_0 the longer they remain arrested complicates the task of identifying the point in G_1 at which arrest occurs.

Temin (492) has studied in some detail the temporal relationship between addition of serum to stationary phase chick cells and the entry into S. He has concluded "that cells were committed to start DNA synthesis about four hours before the actual start of DNA synthesis." It is clear, however, that this commitment in chick cells is reversible, since deprivation of serum coupled with a drop in pH from 7.4 to 6.8 prevents the so-called committed cells from entering S (416).

Cells arrested in G_1 by density-dependent inhibition can also be released by treatment with pronase for as little as 5 minutes (346). The brief exposure to pronase probably causes digestion of protein in the plasma membrane, setting off a train of events beginning with an immediate drop in cyclic AMP (see Chapter 10) and leading to the initiation of DNA synthesis many hours later. These several observations are consistent with the idea of a checkpoint in early G_1 at which cells may be arrested as a result of density-dependent inhibition. They also raise again the question of what events occupy the time interval between this checkpoint and initiation of DNA synthesis.

Pardee's recent experiments (367) on cultured mammalian cells provide further support of the idea of an arrest point in G_1 positioned several hours before the beginning of DNA replication. These experiments show that several different blocking conditions (amino acid deprivation, low serum, elevated intracellular cAMP, and density-dependent inhibition) all act at the same point in G_1. Pardee has introduced the term restriction or R point (restriction point) to describe the block point in G_1.

Reversible Arrest of Tissue Cells in G_1

It is known that some event in G_1 is the focus of the regulation of cell reproduction in tissues. This is based on two observations. First, differentiated cells that cease to reproduce usually contain the G_1 amount of DNA, whether the cessation is reversible (e.g., circulating lymphocytes) or irreversible (e.g., neurons). The remaining cells, containing a G_2 amount of DNA, represent a few tetraploid cells arrested in G_1 and a few diploid cells arrested in G_2. Second, for cells that continue to reproduce in renewing tissues, the rate of reproduction is governed by the average length of time that the cells are retained in the G_1 period. In the mouse, for example, average generation times for proliferating epithelial cells are approximately those listed in the following tabulation (94):

Organ	Generation time (hours)
Esophagus	181
Tongue	40
Duodenum	18.5
Ileum	16.7
Colon	32.6

These wide differences are accounted for almost entirely by changes in the average length of the G_1 period. The average length of the S period in each tissue is about 7 hours, although the average length of the G_2 and mitotic periods tends to increase slightly with increased generation times. As another example of control in G_1, Young (549) found average generation times of 36, 57, and 115 hours in, respectively, the mesenchyme cells of the metaphysis, endosteum, and periosteum of bone, with the differences accounted for almost entirely by differences in the average lengths of the G_1 periods for each population. Similarly in the fetal rat the average generation times for a variety of different cell types ranged from 13.5 hours (intestinal crypt cells) to 40.5 hours (cartilage, parenchymal cells), yet the length of $S + G_2 + M$ remained at the relatively fixed value of 8 to 10 hours (292). For the epithelium of the hamster cheek pouch, $S + G_2 + M$ is about 11.6 hours, while the length of G_1 is 125 hours (73). Blenkinsopp's data (44) on several mouse epithelia indicate that differences in average generation times (from 41 to 8000 hours) are almost exclusively due to changes in the average length of G_1.

In studies on six lines of human lymphocytes in culture with average generation times ranging from 38 to 76 hours, there was some increase in the time taken for mitosis and the S phase in cell lines with longer generation times, but the average length of the G_1 stage was the primary determinant of the length of the cell cycle (18). For unknown reasons the length of the S period in these cells was unusually long, ranging from about 11 to 17 hours.

Finally, analysis of the cell cycles for different populations of cells in the root meristem of *Zea mays* has shown that the main difference between populations is in the average duration of G_1, while the average durations of S and G_2 remain relatively constant (31).

The results cited above are representative of a larger body of published data that have established the following general picture. Changes or differences in the rates of cell reproduction for cells of the same genetic constitution are achieved primarily by expansion or contraction of the average G_1 period. The S, G_2, and D periods may change somewhat; particularly they may increase individually, or all three may increase during the slowing down of the very rapid cell reproduction in embryogenesis (181, 235, 513), but the major basis for regulating the rate of cell reproduction is the retention of cells in G_1. Some confusion about recognition of this fundamental relationship has been introduced by results of studies on various lines of Ehrlich ascites tumor cells. For this cell the length of generation time, G_1, S, G_2, and D may show a variety of relationships depending on the age of the tumor, the age of the host mouse, the sex of the host, and the nutritional state of the host. Ehrlich ascites cells growing in the peritoneal cavity, however, appear to represent an exceptional situation since their behavior does not conform to the behavior of cells in general.

Chalones. At least part of the G_1-arrest mechanism by which cell reproduction is regulated in tissues is probably based on diffusible, negative feedback inhibitors that have been called chalones (chalone = to brake). Chalone activity (measured as inhibition of cell reproduction) has been demonstrated in crude extracts from a variety of animal tissues. The subject of chalones has recently been clearly reviewed by Houck and Hennings (236). Monograph 38 of the

National Cancer Institute (Publication No. 73-425) of the U.S. Department of Health, Education and Welfare (HEW), entitled "Chalones: Concepts and Current Researches" was published in 1973; this volume provides a comprehensive review of the topic of chalones.

The key observation about chalone activity is its tissue specificity. Thus, for example, extracts from epidermis inhibit cell proliferation in epidermis but have no inhibitory effect when applied to other tissues (81). Chalone activity has been reported to be water-soluble, nondialyzable, heat-labile, and precipitable by ethanol, suggesting that the active molecule may be protein (80, 205). Liver cell chalone appears to be a low molecular weight polypeptide (517). Limited evidence suggests that lymphocyte chalone is a glycoprotein (see 236). In most cases the chalones inhibit cells in the G$_1$ period, although both G$_1$ and G$_2$ chalones have been described for epidermis (127, 128). Purified chalone isolated from rat liver prevents hepatocytes from entering DNA synthesis, but has little effect on cells already in S (451). The chalone is specific for hepatocytes, with no significant effect on intestinal or tongue epithelium. Lymphocyte chalone inhibits proliferation of leukemic and lymphoma lymphocytes when added to the medium *in vitro,* but interestingly, the tumor cells are two to four times less sensitive to the chalone in comparison with normal diploid lymphocytes (236). This presumably means that the lesion in the tumor cells involves decreased sensitivity to G$_1$ arrest by the chalone rather than a failure of the tumor cells to produce the chalone.

Prolonged Arrest of Tissue Cells in G$_1$ (Entry into the G$_0$ State)

An extension of the generalization that the rate of cell reproduction is governed primarily by the average length of the G$_1$ period is provided by the observations that the complete cessation of cell reproduction in tissues is achieved predominantly by the prolonged arrest of cells in the G$_1$ period (or entry in G$_0$). Thus, cells that reproduce extremely slowly (in kidney, liver, pancreas, most smooth muscle, etc.) or the cells in nonrenewing tissues (neurons, skeletal muscle cells) all contain the G$_1$ amount of DNA. This is easily demonstrated with the small lymphocytes of peripheral blood. These cells ordinarily do not reproduce *in vivo,* but can be stimulated to reproduce in culture. Every lymphocyte stimulated to divide goes through an S and a G$_2$ period prior to mitosis (394).

Another particularly clear example of the G$_1$ arrest is found in plant embryos. Within a dormant seed the cells of the plant embryo remain in the G$_1$ state. This has been shown by cytophotometric measurement of DNA content (26, 75, 76) and by autoradiographic studies of DNA synthesis in germinating seeds (77, 114, 242).

While it is true that in most plant and animal tissues that have been studied the nonreproducing cells are arrested in G_1 (diverted into G_0), there are exceptions. Gelfant [see (370) for a key to Gelfant's earlier work] has provided extensive evidence that a small proportion of cells in a variety of animal tissues may be arrested in the G_2 stage. In the plant *Vicia faba,* DNA labeling (autoradiographic detection) of germinating seeds has shown that 10% of plant embryos contain a small fraction of cells arrested in G_2 (114). In their studies on germinating seeds of *Pinus pinea* and *Lactuca sativa,* Brunori and D'Amato (75) found only a single embryo with G_2 cells, and these were a minor fraction. The phenomenon of G_2 arrest is discussed further in the section on the G_2 period.

The many observations on plant and animal tissues *in vivo* parallel those for cells in culture, demonstrating that the temporary or permanent restriction of cell reproduction in plant and animal tissues is brought about primarily by the interruption of progress through the G_1 part of the cell cycle.

Absence of the G_1 Period in Some Cells

Any hypotheses designed to explain the presence and significance of the G_1 period must also take into account those situations in which the G_1 period is absent. G_1-less cycles have been observed in unicellular organisms, in certain cells within multicellular organisms (both normal and tumorous), and in cells from multicellular organisms maintained in culture.

No G_1 period is detectable in the slime mold (*Physarum*) (350), in a yeast (*Schizosaccharomyces pombe*) (58), in *Amoeba proteus* (356, 412), or in the micronucleus of two ciliated protozoa [*Tetrahymena* (309); *Euplotes* (256)]. On the other hand the macronucleus in ciliated protozoa (*Tetrahymena, Paramecium, Stentor,* and *Euplotes*) does have a well-defined G_1 period. For *Tetrahymena* and *Euplotes* the cessation of cell division under unfavorable culture conditions results in a G_1 arrest for the macronucleus, but where the micronucleus comes to rest is not known.

The Cell Cycle of Amoeba proteus.   A primary question about cells that do not have a G_1 period is where in the cycle do cells arrest when cell reproduction ceases. In *A. proteus* the initiation of DNA synthesis is tightly coupled to the end of mitosis, and so far no experimental treatments that inhibit cell reproduction have succeeded in uncoupling the two events (Fig. 34). For example, actinomycin D inhibits both cell division and the subsequent initiation of DNA synthesis when administered two or more hours before mitosis (Prescott and Rao, unpublished). It has not been possible to find a point in late G_2 after which mitosis is no longer blocked by actinomycin D, but at which the drug still blocks the subsequent initiation of DNA synthesis. At all time points in late G_2, actinomycin D blocks both subsequent mitosis and DNA synthesis, or it blocks

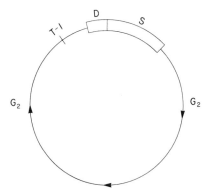

Fig. 34. The diagram gives a summary of the main events that compose the cell cycle in *Amoeba proteus*. DNA synthesis (S period) begins at the end of mitosis (no G_1 period). Most of the cell cycle is occupied by the G_2 period. Two hours before mitosis (in G_2), the amoeba passes T-1 (transition point one). Once the T-1 point is passed, the amoeba becomes insensitive to virtually complete inhibition of RNA synthesis by actinomycin D with respect to reaching mitosis on schedule and initiating the next S period.

neither. The data indicate, among other things, that the decision to initiate DNA synthesis in this G_1-less cell is made 2 hours back in the G_2 period.

Similarly, in the slime mold, which also lacks a G_1 period (Fig. 14), the inhibition of protein synthesis with cycloheximide in early prophase prevents mitosis (111a); however, inhibition of protein synthesis beginning in late prophase allows mitosis to occur and allows the initiation of the S period (although only 20–30% of the DNA is replicated). Evidently the synthesis of any proteins needed to initiate DNA synthesis is completed in prophase and hence some minutes before DNA synthesis is scheduled to begin.

In contrast to cell types that have a G_1 period, it appears that the cessation of cell reproduction in *Amoeba* occurs by an arrest in the G_2 period. The situation, however, is complex. If an amoeba is arrested in the G_2 period (by starving the cell), the subsequent refeeding of the cell does not lead simply to completion of the G_2 period and mitosis. Instead, 2 hours after refeeding, the amoeba initiates new DNA synthesis, which is then followed by a second G_2 period and finally by mitosis. The amoeba can be arrested again in this second G_2 period, and upon refeeding, it will undergo yet more DNA synthesis. The nature of the DNA synthesized under these conditions and its fate during subsequent cell proliferation have not yet been determined. It is evident, however, that the apparent G_2 arrest under starvation conditions is not a simple situation. Under starvation conditions the amoeba probably completes some of the G_2 period and stops in a state that allows the ready initiation of DNA replication upon refeeding. The cell apparently does not progress very far toward mitosis during starvation because, upon refeeding, it proceeds within 2 hours into synthesis of extra DNA and only finally arrives at mitosis much later.

One might argue that those events that normally occur in the G_1 period in most cell types are positioned in the latter part of the G_2 period in the amoeba and slime mold. Viewed in another way, the temporal position of mitosis has been shifted forward within the cycle so that the usual G_1 events now take place during the last hours of G_2.

G_1-less Cells in Multicellular Organisms. A G_1 period is absent in very rapidly proliferating cells within multicellular organisms. This is particularly evident during early embryogenesis. The cleavage stages of sea urchin embryos (144, 230), *Xenopus* embryos (181), snail embryos (513), and mouse embryos (162) lack a G_1 period. For cells that continue rapid proliferation, the G_1-less state persists into later development, for example, the neuroblasts in the grasshopper embryo (164). In most tissues a G_1 period is introduced in the course of differentiation and development, and finally, in the adult, each type settles into some particular average G_1 length. Again, the lengths of S, G_2, and D frequently also increase as the rate of cell proliferation slows during development [see, for example, the results and discussion in Solter *et al.* (468) or Kauffmann (250)], but the primary change is in the average length of the G_1 period.

In the epithelium of hydra the cell cycle proceeds without a G_1 period; the S period is 12 to 15 hours (20–21°C); mitosis lasts 1.5 hours; and the G_2 period is exceptionally long, 24 to 72 hours (113). In an adult mammal the only normal cells reported to lack a G_1 period are rapidly proliferating cells of the myeloid-erythroid series (9). Neoplastic cells, in general, have cycles with G_1 periods that are shorter than their normal counterparts. Under certain conditions, some lines of Ehrlich ascites tumor cells proliferate in the peritoneal cavity of the mouse without a detectable G_1 period (34, 282).

Cultured Cells Lacking a G_1 Period. G_1-less cycles have been observed in two kinds of mammalian cells in culture. For Syrian hamster fibroblasts, the stimulation of arrested cells with serum subsequently leads to at least one cell cycle that lacks a G_1 period (84). Robbins and Scharff (409) have described the cell cycle for a line of Chinese hamster cells (V79) that completely lacks a G_1 period. Apparently, the G_1-less state in V79 cells holds only for cell cultures in log growth. In overgrown monolayers a G_1 period appears, probably because of a slowing down in the rate of cell proliferation. This cell line has potential usefulness for answering a number of questions about the significance of the G_1 period and about the control of the initiation of DNA synthesis. For example, during mitosis the synthesis of all classes of RNA except 4 S and 5 S RNA (560) stops, and the rate of protein synthesis drops by 75%. In the G_1-less hamster cell line, DNA synthesis begins even before the end of telophase, at which time RNA synthesis has only barely resumed, and the rate of protein synthesis has not yet begun to rise. Is the initiation of DNA synthesis independent of the telophase RNA synthesis and of the protein synthesis of the mitotic stages? If so, it might be possible to identify a point in late G_2 or prophase subsequent to which RNA and protein synthesis are no longer needed for the initiation of DNA synthesis in the following late telophase.

As suggested in the case of amoeba, the absence of a G_1 period in any cell type probably means that the events leading up to DNA synthesis do not necessarily have to follow mitosis but may precede it, i.e., the events leading to DNA synthesis may be located in G_2. In this connection it is perhaps important to remember that successive rounds of DNA synthesis and chromosomal duplication can occur without the intervening events of mitosis and cytokinesis. Such an event occurs in the development of polyteny in certain insects, plants, and protozoans. In mammalian cells in culture, two or even three successive chromosome duplications sometimes occur without intervening mitoses (endoreduplication) with the result that the chromosomes are composed of four or eight chromatids at the subsequent metaphase. The molecular basis for this apparent disruption of the normal course of cycle events is not known, but the effect is observed far more frequently following irradiation of cells or following prolonged treatment with fluorodeoxyuridine or amethopterin.

The absence of a G_1 period in the cycles for some mammalian cells growing at maximum rates raises the question of why the G_1 period should continue to be present in any cultured cells growing in rich media. Such media presumably lack any inhibitors that would interrupt the cycle in such a way as to create a G_1 period. One is inclined to make the rather vague assumption that even cells freed of any environmental restriction on growth must overcome in each cell cycle an inherent tendency to be transiently blocked in G_1.

In any case, the existence of mammalian cells that lack a G_1 period shows that the G_1 interval is completely expendable as a stretch of time within the cell cycle.

REQUIREMENTS FOR PROTEIN AND RNA
SYNTHESIS TO COMPLETE G_1

Inhibitors of Protein and RNA Synthesis

There have been several dozen published reports dealing with many kinds of cells, which conclude that G_1 cannot be completed and DNA replication cannot be initiated if a major part of protein or RNA synthesis is inhibited. One of the first such reports is that of Kishimoto and Lieberman (259), and three of the most recent have been provided by Schneiderman et al. (436), Highfield and Dewey (228), and Hereford and Hartwell (225).

Terasima et al. (493, 494) have described the effect of inhibition of protein synthesis for a fixed interval during early, middle, and late G_1 of L cells synchronized by mitotic selection. In each case, entry into S was delayed by a length of time equal to the interval of puromycin treatment. They concluded that the three subdivisions of G_1 were equivalently sensitive to the delaying effect engendered by inhibition of protein synthesis and that protein synthesis is continuously necessary for progress through G_1. In similar experiments Highfield and Dewey (228) reversibly inhibited protein synthesis in CHO cells with puromycin or cycloheximide for given intervals in G_1. In contrast to Terasima and Yasukawa (494) they observed a delay of entry into S that was greater than the length of the interval of inhibitor treatment. In addition, the later in G_1 the interval of protein synthesis inhibition, the greater the delaying effect on entry into S. In agreement with Terasima and Yasukawa the experiments of Highfield and Dewey suggest the need to synthesize proteins continuously to maintain progress through G_1. It is also possible that, in CHO cells, proteins necessary for G_1 progress are more readily degraded as G_1 proceeds, thereby resulting in the reversion of cells to earlier G_1 positions during inhibition of protein synthesis. This apparent erasure of G_1 progress also occurs during deprivation of an essential amino acid (see below).

Unfortunately, more specific information on the role of protein or RNA synthesis in the progress of a cell through G_1 is limited. Whether the requirement for protein synthesis is a generalized one involving many facets of cellular metabolism or whether it is based on the need for one or a few specific proteins is not known. Experiments with the injection of actinomycin D into mice have led to the interesting conclusion that RNA synthesis necessary for the initiation of the S period in epithelial cells of the stomach is completed 6 to 9 hours prior to the S period (152). In cultured Chinese hamster cells and mouse L cells, blocking ribosomal RNA synthesis starting in early G_1 without blocking synthesis of other species of RNA delays the onset of S but does not prevent traverse through G_1 (132, 407). Blocking ribosomal RNA synthesis in late G_1 has no effect on the time of entry of cells into S (Fig. 35). Thus, production of new ribosomes is necessary for the normal rate of traverse through early G_1, but is not needed for a normal rate of traverse through late G_1. However, when cells in G_0 are stimulated to resume proliferation, ribosomal RNA synthesis is required for the reversal of the G_0 state. Once released cells have left G_0 and begun to progress toward DNA replication (are again in late G_1), synthesis or ribosomal RNA is not needed to enter the S period (132).

Protein synthesis is necessary for the initiation of DNA synthesis but not for the continuation of DNA synthesis in the budding yeast *Saccharomyces cerevisiae* (225, 536). This is in contrast to other eukaryotes which require protein synthesis for both initiation and continuation of DNA synthesis. Hartwell *et al.* (216) have obtained many temperature-sensitive mutants affecting different points in the progression of events through the yeast cell cycle (discussed in Chapter 14), and in one of these the affected gene appears to be specifically

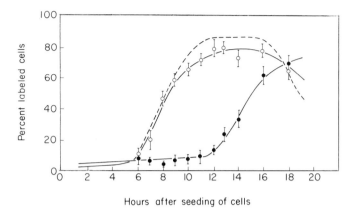

Hours after seeding of cells

Fig. 35. Effect of 0.04 µg/ml of actinomycin D on entry of mitotically selected L cells into DNA synthesis. Nucleolar (ribosomal) RNA synthesis is blocked by 0.04 µg/ml of actinomycin D with no measurable effect on the rest of RNA synthesis. Treatment with actinomycin D for the first 2 hours of G_1 causes a marked delay of entry into S (closed circles). Treatment in later G_1 (3–5 hours after mitosis) does not delay entry in S (open circles). The dashed line shows the normal course of entry into S for untreated cells. Redrawn from Rickinson (407).

needed for the initiation but not the continuation of DNA synthesis. Such genetic approaches offer some hope that the specific protein(s) required for the initiation of DNA synthesis may ultimately be identified in eukaryotes.

Deprivation of Amino Acids

Transit through the G_1 period in mammalian cells has been shown to be much more sensitive to deprivation of an essential amino acid than transit through S, G_2, or D. For example, when the level of isoleucine (498), leucine (138), or tryptophan (74) in the medium is reduced to a trace amount for mammalian cells in log phase growth, all cells in S, G_2, and D are able to finish the cell cycle to become arrested in G_1. Complete removal of an essential amino acid from the medium causes cells to stop at many points scattered throughout the entire cycle (74, 137, 138). Cells in G_1 at the time of partial deprivation of an amino acid are unable to finish G_1. This applies even to those cells that are within minutes of reaching the G_1–S border, as shown by a double-labeling experiment (137). Chinese hamster ovary cells were pulse labeled with ^{14}C-thymidine to label all cells currently in the S period. Immediately thereafter, the cells were deprived of leucine and, at the same time, given ^{3}H-thymidine. Since the pool of thymidine phosphates is sufficient for less than 5 minutes of DNA synthesis, essentially all cells entering DNA synthesis after leucine deprivation will be labeled only with ^{3}H-thymidine. Approximately 50% of the cells were doubly labeled with ^{14}C and ^{3}H, representing cells already in S at the time of leucine deprivation. Less than 1% of the cells were labeled with ^{3}H-thymidine only, which indicates that only cells in the final minutes of G_1 are able to complete G_1 and enter S after deprivation of leucine.

Judging from the time (4 hours) required for the first cells to reach DNA synthesis after readdition of the missing amino acid, it appears that the progress of cells in middle to late G_1 is erased by amino acid deprivation so that such cells revert to an earlier position in G_1. It is possible that the cells revert to the same point to which the S, G_2, and D cells eventually proceed and become arrested as a consequence of partial deprivation of an amino acid.

Amino acid deprivation presumably exerts its effects through a reduction in the rate of protein synthesis. However, during the first 2 hours of isoleucine deprivation for CHO, during which time the amount of DNA synthesis in a log phase culture drops drastically, the rate of protein synthesis measured by incorporation of radioactive amino acids in polysomes was not detectably reduced (129). By 30 hours of isoleucine deficiency, the rate of protein synthesis dropped by 30–40%, and by 48 hours it dropped by 60%. Cells in S and G_2 progress very slowly during amino acid deprivation, and the usual duration of amino acid deficiency needed to accumulate all CHO cells in G_1 is 36 hours. Hence, it appears that traverse through G_1 is sensitive to only partial inhibition of protein synthesis and is far more sensitive to a decrease in protein synthesis than is traverse through S, G_2, and D. The observation that several hours elapse after readdition of the missing amino acid before the first cells enter S may be interpreted to mean that the block point is located in the earlier part of G_1, but it is not really known whether the recovery period represents traverse through a normal set of G_1 events or is concerned with reversal of an abnormal state induced by the amino acid deprivation.

PREPARATION FOR DNA SYNTHESIS

Synthesis of Deoxynucleoside Triphosphates

It has sometimes been suggested that the final hours of G_1 are necessary to synthesize enzymes that function in the provision of deoxynucleoside triphosphate precursors of DNA. This has not proved so for those enzymes studied so far. Thymidine kinase activity is low during the G_1 period for cultured mammalian cells and apparently begins to rise at the G_1 to S transition or in early S, but not before the S phase (for example, 5, 71, 293, 482). [Although in the G_1-less cycle of the slime mold, thymidine kinase activity rises at the very end of interphase (65, 424) and therefore shortly before the beginning of DNA synthesis.] Typically, in mammalian cells thymidine kinase activity

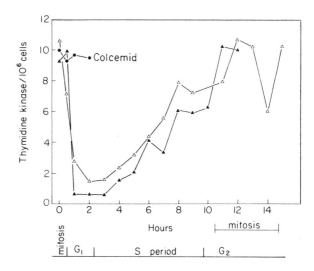

Fig. 36. Changes in thymidine kinase activity during the cell cycle of Chinese hamster fibroblasts synchronized by mitotic selection. The enzyme level falls (two experiments: closed and open triangles) when the mitotically selected cells complete division, remains low during G_1, rises during S, and remains at a maximum level during G_2 and mitosis. Blockage of mitosis with Colcemid prevents the decline in enzyme level (closed circle). Redrawn from Stubblefield and Murphree (482).

reaches a maximum several hours *after* the beginning of DNA synthesis, remains high through G_2, and drops sharply as the cells finish mitosis (Fig. 36) (36, 167, 238, 482, 495). When mitosis is blocked with Colcemid the level of thymidine kinase remains high, at least for an hour after nonblocked cells have entered G_1 and lost most of their kinase activity (167, 482). Thus, completion of mitosis is necessary for the normal program of inactivation or destruction of thymidine kinase in G_1. The maintenance of high thymidine kinase in G_2 and mitosis and its rapid drop at the beginning of G_1 suggest active destruction or inactivation of the enzyme rather than a passive decay process. It would be useful to know whether the enzyme activity decreases in cells in which RNA or protein synthesis are inhibited during the transit from mitosis into G_1.

Inhibition of protein synthesis with cycloheximide or inhibition of RNA synthesis with actinomycin D in G_1 cells abolishes a subsequent rise in enzyme activity, which may mean that the enzyme molecules are destroyed and resynthesized each cell cycle. Cycloheximide and actinomycin D, of course, also block transit from G_1 into S, and DNA synthesis could be causally involved in activation of preexisting enzyme molecules.

Thymidylate synthetase activity in cultured hamster cells is low in G_1 and rises at the beginning of S (108). Similar results have been described for ribonucleotide reductase activity (298, 338), thymidylate kinase (246), deoxycytidine deaminase (168), deoxycytidine kinase (238), and for the synthesis of deoxycytidylate (6).

An explanation for the correlation between the beginning of DNA synthesis and the rise in activity of enzymes concerned with precursor production is suggested by the study of Moore and Hurlbert (329) on the reduction of nucleotides to deoxynucleotides in Novikoff ascites tumor cells. The deoxynucleoside triphosphates of thymidine, uridine, guanosine, and adenosine inhibit the reduction of cytidine and uridine nucleotides. Reduction of GDP is inhibited by dGTP and dATP, and the reduction of ADP is inhibited by dATP. It is plausible that the depletion of deoxynucleoside triphosphate pools at the start of DNA synthesis may release reductases and other enzymes from inhibition as well as derepress their synthesis.

The sizes of deoxynucleoside triphosphate pools varies during the cycle. In Chinese hamster cells the pools of dATP, dGTP, and dTTP are largest during mitosis, and the dCTP pool is largest in late S-early G_2 (519). The triphosphates are quickly degraded after mitosis, and G_1 cells have only a small amount of these precursors (457, 519). The pools begin to expand near the G_1 –S border. In *Tetrahymena* the deoxynucleoside triphosphate pools are largest during the S period, although measurable amounts are present through the entire cycle (341).

DNA Polymerase

DNA polymerase activity shows no particular periodic relationship to the initiation of the S phase (291, 433, 437) or else an increase proceeds in parallel with DNA synthesis but does not precede it (156, 238, 293, 299, 458). Spardi and Weissbach (470) measured the activities of the three different DNA polymerases of HeLa cells in relation to the cell cycle. R-DNA polymerase (γ-DNA polymerase; uses RNA as a template) activity begins increasing near the G_1 –S border. D-DNA polymerase II (high molecular weight enzyme) activity in the cytoplasm increases slowly during the S period, and reaches a maximum after S. D-DNA polymerase II in the nucleus and D-DNA polymerase I (low molecular weight enzyme) in the cytoplasm do not vary in relation to S.

Cleavage synchrony has been used to study the shift of preformed DNA polymerase from the cytoplasm to the nucleus in relation to the cell cycle and in relation to the successive replications of DNA during early development (143). In the unfertilized egg most of the DNA polymerase is located in the cytoplasm. The total amount of polymerase per egg does not change during the first 16 hours of development, but progressively more of this polymerase is found in the nuclei; correspondingly less is found in the cytoplasm with successive cell divisions following fertilization. By the 200–400 cell stage, up to 95% of the DNA polymerase activity is located in the nuclei (144, 295).

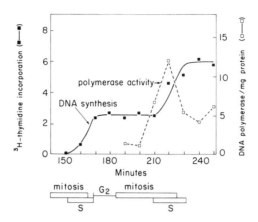

Fig. 37. Relation between DNA synthesis (^{3}H-thymidine incorporation) and DNA polymerase activity in nuclei isolated from 4- and 8-cell-stage sea urchin embryos. DNA synthesis begins in telophase (there is no G_1 period) and is completed in early interphase. Redrawn from Fansler and Loeb (143). Sections of the cell are approximated from Fansler and Loeb (144) and Hinegardner *et al.* (230).

While the polymerase progressively shifts from cytoplasm to nucleus during development, there is also an oscillation in polymerase location during each cycle. Figure 37 shows the nuclear acquisition of DNA polymerase from the cytoplasm as the 4-cell embryos divide and enter the 8-cell stage. A G_1 period is absent in the early cleavage cycles, and DNA synthesis begins in telophase. As DNA synthesis begins in telophase at the start of the 8-cell stage, the amount of DNA polymerase in the nuclei rises sharply. Toward the end of the S period, the nuclear content of DNA polymerase falls and remains low through the subsequent G_2 period and mitosis. Labeling experiments with ^{3}H-leucine show that the cell cycle oscillation in polymerase activity is not due to synthesis and degradation of enzyme but reflects the migration of enzyme molecules between nucleus and cytoplasm in relation to DNA replication (294). The significance of the shift in enzyme location is clear, but what controls the shift is an important unanswered question.

Most of the DNA polymerase accumulated by nuclei during egg development in sea urchins is synthesized and stored in the cytoplasm of the unfertilized egg. Loeb and Fansler (294) have suggested that DNA polymerase may be representative of a general situation in which most of the enzymatic and structural components necessary for rapid cell reproduction are formed and stored in the unfertilized egg.

In summary, G_1 cannot be explained by any known requirements for DNA replication, and the time-occupying events that precede DNA replication remain to be identified. Several kinds of evidence indicate the existence of a checkpoint or arrest point (R point of Pardee; "Start" in yeast) in G_1. It is reasonable to propose that the various situations in which G_1 arrest occurs (regulation of cell reproduction in tissues, density-dependent inhibition of growth, depletion of an essential nutrient, etc.) all involve interruption of the cell cycle at the same checkpoint. The checkpoint may be based on a gene that is activated by a range of both specific and nonspecific environmental signals and prevents advancement toward DNA synthesis.

5

Initiation of the S Period

A main difficulty in understanding the G_1 to S transition is the inability to determine how the replication of the DNA duplexes is initiated in either prokaryotic or eukaryotic chromosomes. Some facts available about the initiation of DNA synthesis in eukaryotes have been mentioned in previous sections. Protein synthesis is necessary for the cell to transit the G_1 period, which is about the only evidence that the initiator protein concept for DNA replication in prokaryotes may operate in eukaryotes. Whether the synthesis of one or more proteins is needed up to the very instant of the initiation of S is not known. The idea that the final event of G_1 is the synthesis of a protein required for initiating DNA replication has sometimes been discussed, but no evidence has been produced. As already mentioned, experiments on whole animals suggest that RNA synthesis is not needed in order to traverse the last part of G_1 (152).

ROLE OF NUCLEAR–CYTOPLASMIC INTERACTIONS IN DNA SYNTHESIS

DNA Synthesis in Cells with Two or More Nuclei

When two or more nuclei share the same cytoplasm, the nuclei almost always proceed through the cycle in extremely close, if not perfect, synchrony. The thousands of physically independent nuclei in the multinucleate plasmodia of the slime mold *Physarum* all initiate DNA synthesis together (350). The same is true for the nuclei in the multinucleated amoeba *Pelomyxa* and for the 50 to 100 macronuclei in the ciliate *Urostyla*. In organisms that ordinarily consist only of uninucleated cells, binucleated cells sometimes arise. For example, in the ciliated protozoan *Euplotes,* organisms with two macronuclei can be found, in which case both macronuclei begin S in precise synchrony (256). Church's results (99) on spontaneously formed binucleated cells in cultures initiated from mouse embryos were substantially the same. The point has, in fact, been well substantiated by observations on a variety of other cell types. Synchrony in the initiation of DNA synthesis in binucleated mammalian cells in culture has, for example, become a commonplace observation in experiments using cytochalasin B. The drug prevents cytokinesis without interfering with mitosis, and the percentage of binucleated cells in a culture approaches 100% as the duration of drug treatment is extended. In our own experience with thousands of such binucleated L cells or Chinese hamster cells, we have always observed that nuclei in binucleated cells are both unlabeled or both labeled (and labeled to the same autoradiographic intensity) following a pulse of ^{3}H-thymidine. Finally, in multi-nucleated cells induced by caffeine in onion root cells, the nuclei in a single cell initiate S synchronously (178).

In a study of cytochalasin-produced binucleates of BHK cells, Fournier and Pardee (148) (in which they confirmed the close synchrony in initiation of DNA synthesis in the two nuclei) discovered that the average G_1 period of binucleated cells was shorter than for the average uninucleated cells. Apparently, in binucleated cells the length of G_1 is determined by the more rapidly advancing nucleus. Initiation of DNA synthesis in the first nucleus evidently triggers DNA synthesis in the second. Whatever the mechanism, it must involve signals that are present in the cytoplasm or pass from nucleus to nucleus through the cytoplasm.

DNA Synthesis in Fused Cells

More striking than these cases of synchrony with nuclei of the same species in a common cytoplasm are the results of Johnson and Harris (249) on heterokaryons consisting of HeLa cells fused with chick erythrocytes. Not only are erythrocyte nuclei (which are normally arrested permanently in G_1) stimulated by the HeLa cytoplasm to synthesize DNA, but the erythrocyte and HeLa nuclei synthesize DNA in synchrony in most cases. The few cases in which an erythrocyte nucleus was labeled with ^{3}H-thymidine but the HeLa nucleus was not were probably the result of continuation of DNA synthesis in the erythrocyte nucleus after the HeLa nucleus had terminated S (asynchronous termination). Among other things the experiments also demonstrate that the cytoplasmic condition that stimulates (regulates?) DNA synthesis is not species-specific.

For the two nuclei in HeLa-hamster and HeLa-mouse heterokaryons, initiation of DNA synthesis "occurred apparently synchronously in the nuclei" and "at a time corresponding to the shorter G_1 period of the mouse or hamster parent, several hours before mononucleate HeLa cells entered the S phase" (186). The two nuclei in a heterokaryon retained their S phase durations and hence terminated DNA replication at different times. The same conclusions were reached by observing DNA replication in the chromosomes of a mouse–hamster hybrid (187). Thus, it appears that DNA replication is initiated by factors or conditions present in the cytoplasm (or in both nucleus and cytoplasm) that are not species-specific. The execution of the S period, however, appears to be guided by a program of events that is intrinsic to the nucleus.

In apparent contradiction to all of the foregoing observations, Sandberg *et al.* (428) found occasional binucleated configurations in a line of leukemia cells in which the two nuclei were clearly out of synchrony with respect to DNA synthesis. The nuclei in bi- and multinucleated Ehrlich ascites cells also sometimes replicate DNA out of synchrony, although synchrony of S periods is the usual rule (86). An explanation of these exceptions is provided perhaps in the experiments on onion root cells mentioned above. While all the nuclei in a multinucleated onion root cell initiate S synchronously, some nuclei finish before others. Thus, as in the heterokaryons discussed above, DNA synthesis is initiated synchronously in nuclei sharing a common cytoplasm, but the programs of synthesis in the several nuclei are not synchronized by nuclear–cytoplasmic interactions even though all of the nuclei are of the same species, and some nuclei finish DNA synthesis before others. Thus, the S period, once initiated, is carried out independently in each nucleus, not only in the case of heterokaryons, but for homokaryons as well. The asynchronous termination of S periods could produce cells in which one nucleus becomes labeled with ^{3}H-thymidine while another does not.

DNA Synthesis in the Micronucleus and Macronucleus of Ciliated Protozoans

Another exception to the usual synchronous behavior of two nuclei in the same cytoplasm occurs in some ciliated protozoa. In *Tetrahymena* (309, 310) and in *Euplotes* (256) the micronucleus replicates its DNA beginning in telophase (and hence a G_1 period is absent), and the macronucleus synthesizes its DNA beginning several hours after micronuclear DNA synthesis has ended (Fig. 38). Why or how the cell achieves temporally different regulations of DNA replication in the two nuclei (which in *Tetrahymena* are believed to be genetically identical) remains an enigma. Micro- and macronuclei differ from each other in a number of ways. In contrast to the macronucleus, the micronucleus contains only tightly packed chromatin (heterochromatin), lacks nucleoli, synthesizes no more than a trace of RNA during the cell cycle, and divides by mitosis (the macronucleus divides by amitosis). It is not apparent how any of these differences might be related to the different positions of the S periods for the two kinds of nuclei. In *Paramecium aurelia,* however, the macro- and micronuclei initiate DNA replication in synchrony (257).

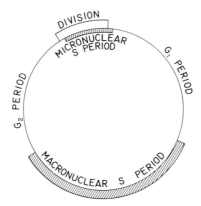

Fig. 38. Diagram of the cell cycle in *Tetrahymena pyriformis,* strain HSM. Micronuclear DNA synthesis occurs in telophase and early interphase. Macronuclear DNA synthesis occupies the middle part of interphase.

Asynchrony of DNA Synthesis in Mitochondria vs the Nucleus

It is also clear that the replications of mitochondrial DNA and nuclear DNA are not synchronous with one another. In the slime mold (200), *Tetrahymena* (92, 96, 368), yeast (537), and chick fibroblasts (312), mitochondrial DNA synthesis has been found to continue throughout the cell cycle. In synchronized mouse lymphoma cells, the rate of mitochondrial DNA synthesis appears to be low in G_1 and early S, high in mid-S, and highest in G_2 (51). A similar pattern has been observed in HeLa cells (379). In a line of liver cells synchronized by cold shock, mitochondrial DNA synthesis was highest during G_2 and D (265). In Chinese hamster cells synchronized by mitotic selection, mitochondrial DNA synthesis occurred between 4 and 13 hours after mitosis, that is, in parallel with nuclear DNA synthesis (287). In cells synchronized by isoleucine deprivation, mito-chondrial DNA synthesis was somewhat synchronized, occurring in the 3-hour interval of 9–12 hours after addition of isoleucine. In regenerating rat liver, incorporation of ^{3}H-thymidine into mitochondrial DNA increases immediately after partial hepatectomy, which is many hours before the initiation of nuclear DNA synthesis (212). It also appears from these studies that the mitochondria in an individual cell do not replicate their DNA in synchrony with each other. This indicates that the initiation of DNA synthesis is controlled autonomously within each mitochondrion.

In HeLa cells synchronized by the double thymidine block method, mito-chondrial DNA appears to be synthesized in two peaks, one in the earlier part of S and one in G_2 (518). The authors suggest that the two peaks could represent DNA replications in two genetically different populations of mitochondria.

According to Pica-Mattoccia and Attardi (379) mitochondrial DNA synthesis is continuous throughout the cell cycle of HeLa cells synchronized by a double thymidine block. There is, however, at least one clear case in which mitochondrial DNA synthesis does appear synchronized. In synchronized *Saccharomyces lactis* the rate of mitochondrial DNA synthesis in the culture rises sharply to a maximum as the cell number doubles and then falls again to a mimimum before the beginning of nuclear DNA synthesis (109, 465). In *S. cerevisiae* arrested in G_1 by mating factor α, mitochondrial DNA synthesis continues unabated for at least 6 hours (374). These various studies lead to the negative generalization that there is no necessarily particular or fixed temporal relationship between DNA synthesis in the mitochondria and DNA synthesis in the cell nucleus.

DNA Synthesis in Recombinants from Different Sections of the Cell Cycle

Several experiments on nuclear–cytoplasmic interactions in the regulation of DNA synthesis are based on recombinations between cells (or cell parts) from different parts of the cell cycle. This is relatively simple to do in amoeba by nuclear transplantation. When an S phase nucleus was transplanted into G_2 cytoplasm, DNA synthesis appeared to be turned off in at least some cases (390). G_2 nuclei implanted into an S phase cell in many cases showed incorporation of ^{3}H-thymidine. Ord (355), however, in a repeat of such experiments has failed to find either of the two effects. The reason for the contradiction is not known, and at the present one can only comment that the long G_2 period in amoeba is more complex than the relatively short G_2 period in most cell types. There is some indication (see earlier discussion of amoeba in Chapter 4) that events usually associated with the G_1 period may be contained in the last part of the G_2 period in amoeba. Possibly, therefore only nuclei in a particular part of G_2, for example the latter part of G_2, can respond to initiating signals in the cytoplasm of an S phase cell.

Ord's results on amoeba agree with those of Guttes and Guttes (201) on the slime mold and, in part, with those of Rao and Johnson (402) on HeLa cells. When pieces of plasmodia in G_2 were fused with plasmodia in S, the G_2 nuclei were not stimulated to synthesize more DNA. DNA synthesis in the nuclei of pieces of S phase plasmodia did not appear to decrease following fusion with G_2 plasmodia. The failure of G_2 nuclei to reinitiate when placed in S cytoplasm suggests that the DNA of G_2 chromosomes is in a condition that prevents response to an initiating signal.

In similarly designed experiments with the ciliate *Stentor,* macronuclei of dividing cells or cells in G_1 initiate DNA synthesis when transferred into S phase cells (118). This points to the presence of an initiating factor in the cytoplasm during the S phase. de Terra (118) argues against the possibility that the absence of DNA synthesis in dividing or G_1 cells is due to the presence of an inhibitor because the consequence of fusion between S and G_1 cells of equal size is initiation in the G_1 nucleus and not inhibition of synthesis in the S phase nucleus. On the other hand, an S phase nucleus transferred to a G_1 cell does stop synthesizing DNA, which indicates the existence of some essential contribution of the S phase cytoplasm to the maintenance (as well as initiation) of DNA synthesis.

Further insight into the situation is provided by the heterokaryon experiment of Johnson and Harris (249) mentioned previously and especially the experiments of Graham et al. (182) on the injection of nuclei into unfertilized eggs of *Xenopus*. Nuclei from embryonic endoderm and from adult liver, brain, and blood cells synthesized DNA within 90 minutes of their injection into an unfertilized egg. Since the nuclei of adult liver, brain, and blood cells are virtually all arrested in G_1, their resumption of DNA synthesis in the new environment is a striking demonstration of the presence of an initiating signal(s) for DNA synthesis in the egg cytoplasm. Remarkably, nuclei of mouse liver are also caused to initiate DNA synthesis in the *Xenopus* egg. The lack of species specificity of the cytoplasmic signal(s) in these experiments has been confirmed by the heterokaryon experiments of Johnson and Harris (249) already mentioned. In further experiments on *Xenopus,* Gurdon (194) showed that the factor in egg cytoplasm that induces DNA synthesis in implanted nuclei is absent from immature oocytes, and appears in the egg just after rupture of the envelope of the germinal vesicle at egg maturation. The factor is apparently produced at this time because it cannot be detected in the germinal vesicle or cytoplasm before breakdown of the germinal vesicle. The factor appears at a stage when the chromosomes belonging to the egg itself enter a condensed state of early meiosis and are obviously unresponsive to the initiation factor "until fertilization stimulates them to proceed beyond the second meiotic metaphase" (194).

Rao and Johnson (402) have made an extensive study of the virus-induced fusion between HeLa cells in different parts of the cell cycle, from which a fairly clear picture has emerged. The fusion of G_1 cells with S cells results in induction of DNA synthesis in the G_1 nuclei. The greater the number of S cells fused with a single G_1 cell, the more rapid is the induction of synthesis in the G_1 nucleus. The G_1 component of a heterophasic G_1/S cell does not inhibit DNA synthesis in the S nucleus. The authors conclude from this work "that certain substances which are present in the S component probably migrate into the G_1 nucleus and cause the initiation of DNA synthesis" (402).

In further experiments, G_2 cells were fused with S cells but there was no induction of DNA synthesis in the G_2 nucleus and no detectable interference in the DNA synthesis in the S phase nucleus (402). Therefore, the DNA in the G_2 nucleus must be in a state that cannot react to the inducing agent proven to be present in the S phase cell. Finally, in fusion products with a high ratio of G_2 to

G_1 components, the G_2 component does not interfere with the completely normal transit of the G_1 nucleus into the S phase.

Rao and Johnson reasonably conclude that their experiments reflect the presence of an inducer of DNA synthesis in S phase cells (positive control mechanism), rather than the presence of a repressor of DNA synthesis in G_1 and G_2 (negative control mechanism).

The total body of these experiments on nuclear–cytoplasmic interactions provides strong evidence that an inducer of DNA synthesis is produced at the beginning of the S phase and probably is maintained throughout the S period. The inducer is apparently absent during the rest of the cell cycle, and the DNA of the G_2 period is not responsive to the inducer. The nature of the inducer of DNA synthesis, the mechanism of its action, the control of its synthesis, and the circumstances of its disappearance are all obviously crucial matters. Several kinds of studies on this inducer are certainly feasible, and such studies are perhaps one of the most incisive ways of attacking the problem of initiation of the S phase in eukaryotic cells. For example, Salas and Green (426) have identified a nonhistone protein that has an affinity for DNA and which is apparently synthesized coordinately with DNA during the cell cycle of mouse fibroblasts. Although there is no evidence that this particular protein is directly concerned with DNA synthesis, the experiments illustrate one way by which proteins concerned with regulating DNA synthesis might be isolated.

DNA Synthesis in Cell-Free Systems

An understanding of the transition of a cell from G_1 to S requires analyses of events in molecular terms. Several starts in this direction have been reported. Friedman and Mueller (157) have devised a system to study DNA synthesis in isolated nuclei of HeLa cells. For maximal activity the system requires the usual supporting factors, plus a heat-labile cytoplasmic factor. The data also suggest that the cytoplasmic factor, while stimulating DNA synthesis in nuclei isolated from S phase cells, is not capable of inducing synthesis in nuclei from non-S cells. In a later paper Kumar and Friedman (275) described the presence of a heat-labile activity in S phase cytoplasm of HeLa cells capable of initiating DNA synthesis in a small percentage of isolated G_1 nuclei. Hershey et al. (226) have shown that proteins of the cytoplasmic fraction from HeLa cells are essential for stabilization of a DNA replication system in isolated S phase nuclei. In a similar experimental arrangement, Thompson and McCarthy (497) have demonstrated the presence of a heat-stable factor in the cytoplasm of L cells and ascites cells that can initiate DNA synthesis in nuclei isolated from cells arrested in G_1 (liver and erythrocyte nuclei). The cytoplasm of G_1-arrested cells (liver) lacks the active factor.

Cytoplasmic extracts prepared from eggs and early embryos of *Xenopus* also have been found to initiate DNA replication in nuclei isolated from adult liver (38). Cytoplasmic extracts of oocytes, hatched embryos, or adult tissue were lacking such initiating activity. The nuclei used in these experiments were presumably arrested in G_1 and had entered G_0. The induced DNA replication was detected both by labeling with ^{3}H-thymidine triphosphate and observation of partially replicated DNA molecules (molecules with replication "eyes") by electron microscopy. The amount of replication was extensive, reaching a level equivalent to a 40% increase in DNA after 15 hours of incubation of isolated nuclei in cytoplasmic extracts. The inducing factor(s) in the cytoplasm is made inactive by heating or trypsin treatment and hence is probably a protein; the factor(s) is not one of the three DNA polymerases known for *Xenopus*. How the factor initiates replication is not known; Benbow and Ford (38) suggest that the factor may be a site-specific endonuclease that initiates synthesis by creating breaks in DNA necessary to initiate replication.

Whether the cytoplasmic factors are really cytoplasmic in the intact cell or have leaked out from nuclei during cell disruption cannot be decided from these several experiments. The fact that similar effects on DNA synthesis are obtained in cell fusion or nuclear transplantation studies does argue, however, that factors involved in initiating and maintaining DNA synthesis are present in the cytoplasm of the intact cell. It is likely that cell-free experiments of the above type could provide a much better understanding of the G_1 to S transition and the regulation of DNA synthesis.

INTRANUCLEAR SITE OF INITIATION AND CONTINUATION OF DNA SYNTHESIS

Electron Microscope Autoradiography

Comings and Kakefuda (107) concluded from electron microscope autoradiography of synchronized human amnion cells that DNA replication at the beginning of the S period is initiated at the nuclear envelope. They suggested, in parallel with the apparent situation in prokaryotes (see 274), that replicons (at least those that are triggered at the beginning of S) are attached to membrane at or near their initiation regions. In contradiction, Williams and Ockey (535) found in synchronized Chinese hamster cells that the initial incorporation of ^{3}H-thymidine at the beginning of S was not localized at the nuclear envelope but distributed throughout the nucleus. At later and later stages of S the incorporation of ^{3}H-thymidine was restricted more and more to the periphery of the nucleus, presumably reflecting the late replication of heterochromatin, which strongly tends to be condensed up against the nuclear envelope. Other observations confirm this general picture (45, 133, 251).

The question of an association of DNA replication with the nuclear envelope has recently been reexamined with careful experiments in four different laboratories with cultured mammalian cells using electron microscope autoradiography (106, 140, 240, 541) (Fig. 39). The four reports completely agree that neither the initiation of DNA synthesis at the beginning of the S phase nor the replication forks responsible for the continuation of DNA synthesis are associated with the nuclear envelope. The well-known attachment of DNA to the nuclear envelope in animal cells therefore appears to have nothing to do with the control, initiation, or execution of DNA replication in these cells.

In the ciliate *Euplotes eurystomus,* the macronucleus is a cylinder about 100 μm long and 8 μm in diameter. DNA synthesis takes place in two replication bands that travel through the macronucleus (Fig. 40). A replication band occupies the full diameter of the macronucleus, and DNA is fed into the band as it moves down the nucleus. The DNA molecules in the macronucleus, however, have an average weight of 1.5×10^6 daltons (equal to an average length of 0.75 μm) (386). These molecules replicate throughout the full 8 μm diameter of the replication band (Fig. 41). Hence, those replicating molecules contained in the inner part of a band cannot possibly be attached to the nuclear envelope because they are not long enough.

DNA Synthesis and Isolated Nuclear Membranes

Using the M-band technique developed by Tremblay *et al.* (509) to isolate DNA-membrane complexes of bacteria in sucrose gradients, several laboratories have reported that in eukaryote cell lysates, newly replicated DNA is associated with a nuclear membrane fraction (207, 251, 326, 369). An M-band (M for membranes) is formed by mixing isolated nuclei with the ionic detergent, sarkosyl, in the presence of magnesium ions and then centrifuging in a sucrose gradient. The sarkosyl and magnesium form crystals that in turn form a complex with membranes, probably by hydrophobic interaction. The crystal with attached membrane sediment to form the M-band. In bacteria the M-band contains more than 90% of the DNA, if precautions are taken not to shear the DNA. This result is taken as evidence that in bacteria one or more points along the chromosome are attached to the plasma membrane. Pulse-labeling experiments indicate that these attachment points are the origin of DNA replication and the replication forks.

The results of M-band experiments with eukaryotes lead to the conclusion that the DNA replication machinery is attached to the nuclear envelope, a conclusion clearly contrary to the autoradiographic results cited above. It is possible that the DNA-membrane association observed in the fractionation studies is an

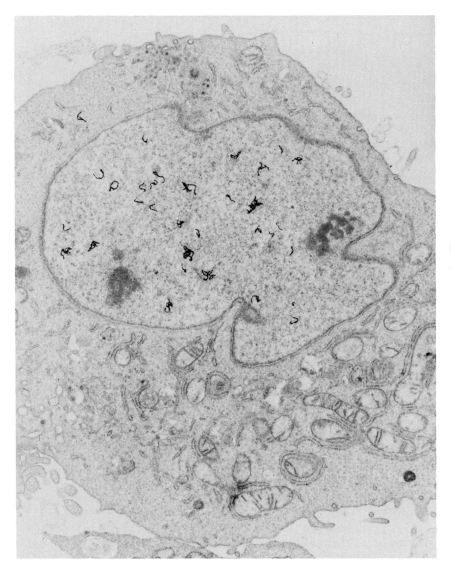

Fig. 39. An electron microscope autoradiograph of a hamster cell labeled with ³H-thymidine for 5 minutes at the beginning of the S period. The silver grains are located over the more central regions of the nucleus showing that DNA synthesis is not initiated at the nuclear envelope. From Wise and Prescott (541).

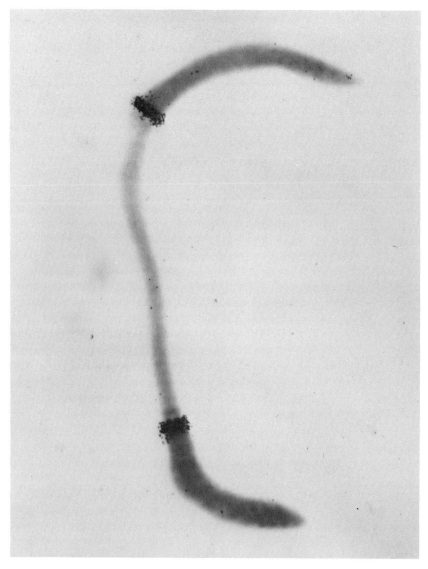

Fig. 40. A macronucleus of *Euplotes eurystomus* isolated and autoradiographed after labeling with ³H-thymidine for 10 minutes. The autoradiographic grains show the positions of the two replication bands, which are moving toward each other.

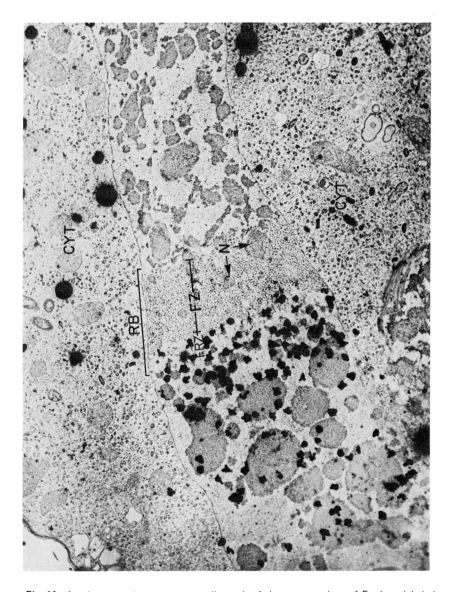

Fig. 41. An electron microscope autoradiograph of the macronucleus of *Euplotes* labeled for several minutes with ³H-thymidine. The presence of newly labeled DNA in the replication band (arrow) several micrometers from the nuclear envelope proves that at least the bulk of DNA synthesis occurs without association with the nuclear envelope. RB, replication band; FZ, forward zone of replication band; RZ, rear zone of replication band; N, nucleoli; CYT, cytoplasm. From Evenson and Prescott (135). Photomicrograph by A. R. Stevens.

artifact incurred during or after cell or nuclear lysis. The replication complex, which contains native and single-stranded DNA and at least several protein molecules, may have hydrophobic properties that might result in binding to sarkosyl crystals and thus readily lead to such an artifact.

In contradiction to the M-band results, analysis of purified nuclear membranes of liver cells has shown that the DNA bound to the membranes is labeled at a lower rate than bulk DNA except during the late S period (151). This is consistent with the known attachment of heterochromatin to the inner nuclear membrane and the replication of heterochromatin DNA late in S, and therefore rules against attachment of replication origins and forks to the nuclear membrane. This agrees with the observation that newly synthesized DNA is not preferentially associated with nuclear envelopes purified from mouse cells (140).

Also, according to the nuclear fractionation studies of Oppenheim and Wahrman (354), DNA replication does not take place at the nuclear envelope in *Physarum*. The replication of adenovirus DNA in the nuclei of HeLa cells is not associated with the nuclear envelope (452). Finally, fractionation of nuclei from regenerating liver has shown that the DNA most rapidly labeled with ^{3}H-thymidine is associated with a particular structural protein matrix within the nucleus away from the nuclear envelope (41).

In summary, most of the experimental evidence acquired so far strongly rules against the involvement of the nuclear envelope in the initiation of DNA replication and rules against attachment of the DNA replication machinery to the envelope after initiation.

6

The S Period

In prokaryotic cells the entire genome is contained in a single DNA molecule that has the form of a closed loop in some species at least and perhaps in all. Replication of the molecule during the S period begins at a single genetically defined origin (274) and proceeds bidirectionally around the circular molecule (387).

In contrast, eukaryotic cells contain manyfold more DNA that is distributed in two or more chromosomes. In eukaryotes the nuclear DNA molecules are linear and in most cases much larger than in prokaryotes. The circular DNA molecule in *E. coli*, for example, has a molecular weight of about 2.6×10^9 daltons. The two largest chromosomes in *Drosophila melanogaster* each contain a single molecule of DNA with a molecular weight of about 40×10^9 daltons. Chromosomes of mammalian cells vary widely in size, but the average mammalian chromosome contains about 3 cm or 6×10^{10} daltons of DNA. In the salamander *Amphiuma means*, the chromosomes are among the largest known. The average chromosome in *Amphiuma* contains an amount of DNA equivalent to a double helix about 2.5 m long, and this also is apparently all contained in a single molecule.

The greatly increased amount of DNA in eukaryotes compared to prokaryotes and its distribution among many chromosomes, often with enormous DNA molecules, increases the complexity of replicating the genome. In addition, chromosome replication in eukaryotes includes the assembly of many proteins, histones and nonhistones, into association with the DNA. Much has been learned about the way in which eukaryotic chromosomes replicate, although the regulatory mechanisms that provide for the orderly progression of the S period are still largely not understood.

THE NUMBER OF REPLICATING UNITS (REPLICONS)

Replication of DNA in a eukaryotic cell is accomplished by a large number of replicating units or replicons. The presence of multiple replicons in individual eukaryotic chromosomes was first indicated by the autoradiographic studies by Taylor (489) and subsequently by many others. When cells are given a brief pulse of ^{3}H-thymidine (e.g., 10 minutes) during the S period and then examined by autoradiography when they subsequently reach mitosis, radioactivity is detected at many points along individual metaphase chromosomes. Hence, during the 10-minute pulse (plus perhaps a 5-minute chase period) DNA synthesis must have been in progress at many points along individual chromosomes.

Cairns (89) provided the first approximation of the minimum number of replicons in the average chromosome of a mammalian cell by autoradiography of DNA molecules. HeLa cells were labeled with a brief pulse of ^{3}H-thymidine, the DNA isolated, spread, and autoradiographed. The length of grain tracks indicated a rate of replication of 0.5 μm/minute. From this rate and knowing that the average HeLa cell chromosome contains 3 cm of DNA, simple calculations suggested that the average chromosome must have more than 100 separate replication points in order to accommodate replication of the total DNA within the normal S period of 6 to 8 hours. In fact, the grain tracks in this experiment probably represented several replicons, each replicating bidirectionally. This would lead to an overestimate of the rate of travel for a single replication fork. However, the replicative events producing the grain tracks probably began and ended during the ^{3}H-thymidine pulse, producing an underestimate in the rate of fork travel. Hence, the calculation of the minimum number of replication points in the experiment contains a large error. The experiments were important, however, because they introduced the Cairns technique of DNA fiber autoradiography into the study of eukaryotic chromosomes, which later led to major discoveries by Huberman and Riggs (241).

Two other estimates of the number of replicons in eukaryotes were made prior to the experiments of Huberman and Riggs. Plaut *et al.* (381) could identify 30 separate points of simultaneous DNA replication in 15% of the total length of the chromosomes in *Drosophila*. This suggests by extrapolation that the relatively small genome in *Drosophila* {0.18 pg of DNA (see 403) vs 3.0 pg in a mammalian cell [see Laird's review (278) of the available data]} contains at least 200 replicons in its four chromosomes. There is some evidence [see the discussions by Pelling (372) and Mulder *et al.* (336)] that each band in the polytene chromosomes of *Drosophila* is a separate replicating unit, which would put the number of replicons up to about 5000 per genome in the four polytene chromosomes.

By a method designed to measure the number of replication forks in HeLa DNA, Painter *et al.* (364) estimated that 1000 to 10,000 replicons are synthesizing DNA at any one time. The total number of replicons per diploid cell will be considerably higher since most if not all replicating units replicate in less than the time required for the entire S period.

The most accurate estimate of the number of replicons in mammalian cells is provided by the experiments of Huberman and Riggs (241). They labeled DNA in Chinese hamster cells and HeLa cells with short pulses of ^{3}H-thymidine, purified the DNA, and carried out autoradiography. From the length of the autoradiographic patterns it is apparent that most replicons are between 7 and 100 μm long. These results have been confirmed by Lark et al. (283). Assuming that the average replicons are 30 μm long and adding the fact that a haploid set of mammalian chromosomes contains about 90 cm (3 pg or 1.8 × 10^{12} daltons), the number of replicons (30 μm long or about 6 × 10^7 daltons) per haploid genome must be about 30,000. The Chinese hamster lines used in the above autoradiographic experiments have chromosome complements of 21. Hence the average number of replicons per chromosome in this case must be about 2800.

Some of the autoradiographic patterns observed by Huberman and Riggs suggested that DNA replication might proceed in two directions from a single origin. Such bidirectional replication has now been confirmed as the predominant mode of replication for the replicons in animal cells (10, 90, 209, 210, 240, 526).

The time required for bidirectional replication of a 30-μm-long replicon can be calculated from the rate of travel of the single replication fork. From autoradiographs the following rates have been measured: in the HeLa cell, 0.5 μm/minute (89); in the Chinese hamster cell, 0.5 to 1.2 μm/minute (241), 1 to 2 μm/minute (488), and 0.8 μm/minute (283). From a physical–chemical analysis Lehmann and Ormerod (285) calculated a rate of 0.7 to 1.0 μm/minute for a mouse lymphoma cell, and Painter and Schaefer (363) have calculated rates of about 0.5 to 1.1 μm/minute in HeLa cells (Painter and Schaefer showed that the rate is 0.5 μm/minute in early S and 1.1 μm/minute in later S). The average of all these values for mammalian cells is 0.9 μm/minute per fork. Therefore, the time required for replication of the average mammalian replicon (30 μm) is about 17 minutes.

MINIMUM NUMBER OF REPLICON FAMILIES

From the determination of 17 minutes as the time needed to replicate the average replicon, the minimum number of families or banks of replicons necessary to accomplish replication of the genome may be estimated. A bank is defined as a group of replicons that initiate replication simultaneously, presumably because they all respond to the same control signal, but there is no firm evidence that groups of replicons are controlled by a common signal. The autoradiographic studies show that the members of a putative bank are scattered throughout most or all of the chromosome complement, interspersed with members of other banks (see particularly 234). Since the mammalian S period lasts 7 hours, a minimum of 25 banks of replicons initiating in sequence is needed to maintain a continuous high rate of DNA synthesis through the S period. Since a haploid mammalian cell contains about 30,000 replicons, the average bank would contain over 1000 replicons. The real number of these putative banks of replicons may, in fact, be greater than the minimum estimate of 25, and banks may contain different numbers of replicons. It is conceivable, for example, that the DNA in a single satellite component may all belong to a single bank. For such satellite DNA's as those in mouse and kangaroo rat cells (308) the bank of replicons would be quite large since 10% or more of the total cellular DNA can belong to a single satellite.

APPARENT CHANGE IN THE NUMBER OF ORIGINS
OF REPLICATION IN RELATION TO DEVELOPMENT

A complication in determining how the replication of replicons is controlled and ordered in time has emerged from studies of DNA replication in cells of the same species that are in different developmental states. In prokaryotes in general, replication always begins at a particular genetically defined point on the circular DNA molecule. It was more or less tacitly assumed initially that the replicons identified in eukaryotes were also genetically defined units each possessing a single fixed origin, but this has proved not to be so. This is shown by the following situation described by Callan (90).

The S period immediately preceding male meiosis in *Triturus* lasts 9 to 10 days at 16°C (91). Somatic cells of *Triturus* at the same temperature take about half that time for DNA replication. Furthermore, embryonic cells replicate their genomes about 50 times faster than adult somatic cells. These different lengths for the S phase are not the result of different programs of replication of replicons of fixed number and size (greater and greater temporal spacing in initiation of successive replicons in going from embryonic cells to adult somatic cells to meiotic chromosomes). Also, the differences in lengths of the S periods in the three situations are not due to differences in rates of travel of replication forks, i.e., to different times to replicate replicons of fixed size. Instead, the differences in S length are apparently due to *changes in the sizes* of replicons. In the premeiotic chromosomes the number of initiation points is greatly reduced, and the replicons are correspondingly longer and require more time to replicate. At the other extreme, embryonic cells have far more initiation points, hence much shorter replicons and a much shorter S phase.

Another example of this alteration in the size of replicons occurs in *Droso-*

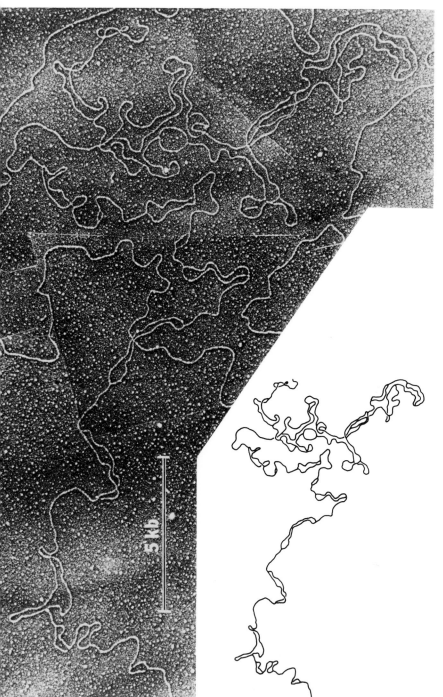

Fig. 42. Replicating DNA from chromosomes of cleavage stage nuclei of *Drosophila* embryos. The portion of the molecule shown is 119 kilobase pairs long (equivalent to about 39 μm or 78 × 10⁶ daltons) and contains 23 regions of replication. Reprinted with permission from Kriegstein and Hogness (270).

phila. The S period in embryonic nuclei is 3 or 4 minutes of a total cycle time of 10 minutes. Electron microscopic observation of replicating DNA from cleavage stages of embryos shows that the genome is divided into replicons that are on the average about 2 or 3 μm long (Fig. 42) (46, 270, 542). Apparently, essentially all these short replicons are initiated within 0.4 minutes of each other at the beginning of S. Therefore, since the rate of fork travel in these nuclei is 0.9 μm/minutes (46), most replicons, replicating bidirectionally from their origins, should complete replication (by meeting of replication forks of adjacent replicons) within the first 2 minutes of the S period. The final length of S will be determined by the largest replicons, which are about 6.5 μm long and therefore require about 4 minutes to replicate.

In marked contrast to the 3–4 minute S period of cells of the early embryo, *Drosophila* cells maintained in cell culture have an average S period of 600 minutes (46), yet the rate of travel for replication forks appears to be the same. To account for the greatly increased lengths of S, the number of active origins of replication (number of replicons in the DNA) must be greatly reduced or the initiation at origins must be spaced out over a hundredfold greater time span. Blumenthal *et al.* (46) suggest that formation of chromomeres during embryogenesis, each containing a large fraction of condensed DNA, is accompanied by inactivation of the origins in the condensed regions. In polytene chromosomes the condensed part of a chromomere becomes a band and accounts for most of the DNA. In the interchromomeric region (interband of polytene chromosomes) the origins can still function in the initiation of replication. The DNA in the regions of inactivated origins (bands) then must replicate by forks arriving from active origins in the interbands. This involves long travel distances for individual forks that could account for the long S period. This idea that replication is accomplished by many fewer forks is supported by the much lower frequency of forks seen by electron microscopy in DNA from cultured cells in S. The hypothesis also must include the assumption that termini of replication are not genetically determined points on the DNA but occur wherever two replication forks generated from successive origins meet each other.

THE ORDERING OF DNA REPLICATION

Maintenance of the Same Order in Successive Cell Cycles

We still do not know much about the temporal ordering of DNA replication in each S period, but what data are available consistently indicate that the ordering is relatively constant from one cell cycle to the next. Several experiments have been done with eukaryotic cells that follow the experimental plan used by Pritchard and Lark (399) who showed that the starting point for replication of the chromosome in *E. coli* was fixed and constant and that replication always proceeded with the same sequential ordering. Braun *et al.* (68) and subsequently Braun and Wili (66) took advantage of the very high degree of natural synchrony of DNA synthesis among the nuclei in the plasmodia of the slime mold. A pulse of ^{14}C-thymidine was given at the beginning of the S period in order to label only the DNA made in early S. In the next cell cycle the plasmodium was incubated in bromodeoxyuridine during the beginning part of S, thereby selectively increasing the density of the DNA made in early S. The experiments clearly demonstrate that the particular DNA synthesized in the first part of S (^{14}C-labeled) in one cell cycle is also replicated in the first part of S (Budrlabeled) in the next cell cycle. In further experiments Braun and Wili (66) divided the S period approximately into fifths and showed that all five subfractions of DNA replicated in the same order from one cycle to the next. Muldoon *et al.* (337) have refined this kind of analysis, dividing the S period of the slime mold into 10 consecutive intervals of DNA synthesis in which the ordering of the intervals is retained from one cycle to the next.

In an experiment of similar design Mueller and Kajiwara (333) showed for HeLa cells that DNA which was replicated early in the S period of one cell cycle was also replicated early in the S period in a subsequent cell cycle. Taylor *et al.* (487) refined the experiment slightly and showed for Chinese hamster cells that both early and very late replicating DNA's retain their temporal positions of replication from one cell cycle to the next. Unfortunately, it is not feasible to refine this kind of analysis to precisely identifiable short segments of the S period (and hence to banks of replicons or parts of replicons) because no method is available for synchronizing cells precisely enough with respect to the S period. The above experiments, nevertheless, might reflect an ordered cascade of replications of replicons in the eukaryotic nucleus.

There is no way at present to decide, once S has started, whether the presumed ordering of replicons occurs because the replication of a given replicon (or bank of replicons) causes the initiation of a subsequent replicon (or bank) or whether the entire sequence is programmed by another mechanism. The latter idea is perhaps more appealing because it could also include the control of the first replicons that are initiated at the beginning of the S period. The experiments of Cummins and Rusch (111a) on the slime mold do suggest that the earlier part of DNA synthesis must be completed before later synthesis can be initiated.

The hypothesis that the sequential order of replicon initiations is determined by the replicons themselves has been tested in a limited way by observing replication patterns of chromosomes in hybrid cells (289, 301). In hamster–mouse hybrids variable numbers of the hamster chromosomes are lost. Yet in some of these hybrids the hamster chromosomes that are retained appear to replicate their various parts in the same order as the parental hamster chromosomes. In other hybrids the replication pattern of the remaining hamster chromosomes appears to have been altered. In spite of the lack of consistent results, these experiments suggest that the entire genome need not be present for at least some chromosomes to retain their original, intrachromosomal, sequential replication patterns. In the face of this evidence that a given chromosome can be autonomous in its replication pattern, it is difficult to see how the order of initiation of banks of replicons could be governed by a sequential interaction of banks of replicons scattered throughout all of the chromosomes of a genome.

Similarly, in human—mouse cell hybrids in which as little as 25% of the human chromosome complement has been retained, the human chromosomes show the same terminal replication sequence observed in the parental cell. Hence, "the maintenance of the normal terminal replication sequence in human chromosomes does not depend on the human genome being intact" (289).

Both experiments, therefore, suggest that the pattern of DNA replication is determined independently within each chromosome.

The GC to AT Shift during the S Period

Further evidence of ordering of DNA replication in mammalian cells is reflected in the differences in buoyant densities of DNA's replicated at different times in S. In HeLa cells synchronized by the mitotic shake-off technique the buoyant density of the DNA is highest for DNA synthesized near the beginning of the S period and declines as the S period proceeds (504). That the buoyant density shift represents a shift in the base composition of the DNA was proved directly by base analysis. DNA replicated early had an average GC content of 43.6% and DNA replicated late in S had an average GC content of 38.7%. The average GC content for DNA pulse-labeled in an asynchronous culture was 40.1% (close to the base composition of the total DNA). Again, using methods to synchronize cells at the beginning of the S period, Tobia et al. (504) and Flamm et al. (146) demonstrated a change in buoyant density (GC to AT shift) during the S period of mouse L cells and mouse lymphoma cells.

The phenomenon of the GC to AT shift can, at present, be studied with somewhat more precision by "retroactive" synchronization (see Chapter 3). With the retroactive method the shift in buoyant density (GC to AT shift) has been shown to occur continuously from the beginning to the end of the S period in L cells and Chinese hamster cells (54, 55) and in an untransformed strain of rabbit euploid cells (56). Thus, the steady shifting from the synthesis of relatively GC rich to relatively AT rich DNA during the S period appears to be a general phenomenon in mammalian cells, occurring in both nontransformed and transformed cell types.

Similarly, in the slime mold, DNA replicated in early S has a higher buoyant density than DNA replicated later in S (64). This shift in base composition is another indication of a high degree of ordering in the sequence of replication of

DNA segments in eukaryotic cells. Beyond this, the significance of the GC to AT shift is not known. The data on rabbit cells indicate that the shift in base composition is not due to the shift from the synthesis of euchromatin early in S to the synthesis of heterochromatin in late S (56).

Replication of Buoyant Density Satellites
and Repetitious DNA's

The occurrence in the nucleus of DNA fractions with different buoyant densities has provided another way to analyze progress in the S period. In a species of the kangaroo rat (*Dipodymus ordii*) a number of density satellites account for well over half of the total nuclear DNA (220). The three satellites present in the largest amounts could be shown to dominate DNA synthesis at three different parts of the S period, but none showed measurable replication at the beginning of the period (53). The S period in this species begins with the synthesis of main band (nonsatellite) DNA.

Tobia *et al.* (503, 504) noted that the AT-rich satellite of mouse cells is mostly synthesized "after main band DNA has reached its maximum rate of synthesis." In L cells synchronized by treatment with fluorodeoxyuridine Cohen *et al.* (105) have also found that satellite DNA replicates late in S. By the more precise method of retroactive analysis of the cycle, the timing of synthesis of satellite DNA has been shown to be restricted to the third quarter of the S period of L cells (54). Smith (464), however, reported that the satellite replicates at the beginning of the S period in mouse cells undergoing transformation with polyoma virus. This could mean that the order of replication of DNA elements may change under certain conditions, but Hatfield and Walker (221) have been unable to verify Smith's report. They conclude that there is no difference in the timing of satellite DNA replication between cells infected with polyoma virus and uninfected cells.

In this same context, however, it should be noted that the replication of the heterochromatic X chromosome in a female rabbit cell begins late in S, but during early embryonic cleavage both X chromosomes initiate replication early in S (258). Similar to the rabbit, the early replication pattern of the sex chromosomes is retained during embryonic cleavage in the hamster (229). Also, half the X chromosome and all the Y chromosome begin replication late in somatic cells of the hamster but change to early initiation during spermatogenesis (335, 512). The same type of shift in replication time occurs in the X chromosome of the grasshopper during meiosis (344). These observations show that the order in which certain DNA segments are replicated may be changed, and the change is associated with a shift from a heterochromatic to a euchromatic state, or vice versa.

In contrast to the replication late in S of the highly repetitious DNA sequences that make up buoyant density satellites, other repetitious sequences in mammalian cells are synthesized throughout the S period (206, 435). In chick cells some repetitious sequences of high buoyant density are synthesized throughout S while repetitious sequences of low buoyant density are synthesized only during late S (485).

Replication of rDNA

In a further example of a defined order within the S period, Amaldi *et al.* (11) determined by RNA–DNA hybridization that the cistrons for ribosomal RNA (rDNA) replicated between 1.5 and 3.0 hours of the S phase in Chinese hamster cells. Stambrook (472) has similarly found that the ribosomal RNA cistrons in Chinese hamster cells replicate early in S, but their replication is apparently spread out over several hours. In rat kangaroo cells the ribosomal cistrons (rDNA) have been reported to replicate late in S (172), but in HeLa cells Balazs and Schildkraut (27) failed to find restriction to a particular part of S of the replication of ribosomal cistrons. Replication of rDNA in *Tetrahymena* occurs in a short interval near the beginning of the macronuclear S period (13). In *Physarum,* the replication of ribosomal cistrons appears to begin about one-third through the S period and to continue long after the rest of nuclear DNA has finished replication, i.e., through the G_2 period (340, 551). These experiments extend the earlier observation that DNA synthesis takes place in nucleoli in *Physarum* during the G_2 period (199). The synthesis of ribosomal cistrons in G_2 is far less sensitive to inhibition of protein synthesis by cycloheximide than is the major nuclear DNA (530). At present no explanation can be offered for the very prolonged period of replication of ribosomal cistrons.

Replication of rDNA also demonstrates that the replication of particular DNA segments can be changed from the usual pattern. Well-documented cases are the differential synthesis of the genes for ribosomal RNA during oogenesis and embryogenesis (see 161, 471). Other examples of differential DNA replication are the exclusion of heterochromatin from DNA replication during polytene chromosome formation in *Diptera* (336, 418) and the selective nonreplication of heterochromatic chromosomes in the mealy bug (349).

To summarize to this point, the S period in eukaryotic cells appears to be composed of a cascade of initiations and replications of the many thousands of replicons in a single nucleus. (But note the replicative behavior of DNA in *Drosophila* embryos.) It is probably simplest to assume that the replicons occur in banks with all members of a bank responding to an initiation signal that is specific for the bank. How the putative signals might be programmed to appear at the appropriate times is probably beyond reasonable speculation at present. It is usually assumed that the program of signals involves protein synthesis, the synthesis of a new protein being required to initiate each successive bank of replicating units. This general concept, however, must be modified to explain how the timing of initiation of replicons can be shifted between early and late S (in parallel with shifts of the replicons between the euchromatic and hetero-chromatic state).

REQUIREMENTS FOR PROTEIN AND RNA SYNTHESIS DURING THE S PERIOD

Protein Synthesis

Many people have reported that the inhibition of protein synthesis (in most cases with cycloheximide or puromycin in mammalian cells) results in the immediate rapid decline in the rate of DNA synthesis. Bennett *et al.* (40) were among the first to describe the effect, and some of the most recent reports are those of Wanka and Moors (522) on *Chlorella* and Ensminger and Tamm (131), Hand and Tamm (209), and Seki and Mueller (444) on mammalian cells.

In a general way, the rapidity of the decline in the rate of DNA synthesis following the inhibition of protein synthesis does fit the idea that the initiation but not the continuation of the replication of a replicon requires new protein synthesis. If the average replicon is 30 μm long and the rate of DNA replication is 0.9 μm/minute (see earlier discussion), then, as replicons terminate, the rate of DNA synthesis in a cell population should decrease by about 50% during the first 10 to 15 minutes of inhibition of protein synthesis and decrease to near zero in about 30 minutes. The inhibition of DNA synthesis that follows the inhibition of protein synthesis is a little slower than this, and also DNA synthesis persists at a low rate for many hours (far beyond the normal length of the S period) after the initial rapid decline (137). This suggests either that a few replicons can initiate during inhibition of protein synthesis or that the rate of traverse of the replication fork is very much slowed in some replicons or that some replicons are very long. Inhibition of protein synthesis with puromycin does not appear to reduce the rate of fork travel (131, 211, 234), although more recent work disputes this (165). Cycloheximide reduces the rate of replication fork travel by about 50% (166, 209, 527). Both Weintraub and Holtzer (527) and Gautschi and Kern (166) conclude that the effect of cycloheximide on DNA synthesis for about the first 2 hours is due to slowing of replication forks and does not involve inhibition of new initiations of replicons. Weintraub (526) has proposed that the slowing of fork travel is a consequence of inhibition by cycloheximide of histone synthesis and that these proteins are needed as chain elongation factors in DNA replication. It would appear, however, that more experiments using other systematic approaches, for example, inhibition of protein synthesis by deprivation of essential amino acids, are needed to clarify the relationship of protein synthesis to DNA synthesis in cultured vertebrate cells.

In *Physarum* the amount of DNA synthesized increases "in discrete steps in response to variation in the time of cycloheximide addition" (337). Ten steps could be resolved, suggesting that the slime mold genome consists of at least 10 banks of replicating units, with the banks "controlled by proteins synthesized at defined times during the S period."

Although the proposed cascade of initiations of replicons during S may reasonably be supposed to be guided by the synthesis of a sequence of inducer (initiator) proteins, no direct evidence for such an idea has been obtained as yet. In addition, the continuation of DNA synthesis may require the continuation of histone synthesis as most recently suggested by Weintraub (526). This idea has been discussed from time to time because of the knowledge that histone synthesis is initiated with and proceeds in parallel with DNA synthesis. It is true that the arrest of DNA synthesis results in the rapid disappearance of histone synthesis, but it is difficult to determine whether histone synthesis is required for continuation of DNA synthesis. Conceivably the problem could be partially studied by inhibiting protein synthesis by means of tryptophan starvation of S phase cells. Such deprivation should sharply depress general protein synthesis but presumably have no direct effect on histone synthesis. If tryptophan deprivation were to have markedly less effect on DNA synthesis than does deprivation of other amino acids, this would imply that histone synthesis is the facet of protein synthesis most immediately required for DNA synthesis. In support of this idea Weintraub (526) has shown that DNA synthesis in erythroblasts is relatively insensitive to analogues of tryptophan, but highly sensitive to canavanine, an analogue of arginine. There is also an indication in the earlier study by Freed and Schatz (154) that tryptophan deprivation has less effect on DNA synthesis in Chinese hamster cells than deprivation of lysine or arginine.

An exception to the generalization that protein synthesis is necessary for continuation of DNA synthesis during the S period in eukaryotes was mentioned earlier in the case of budding yeast (225, 536). In yeast the absence of dependence of DNA synthesis on protein synthesis after the S period has started apparently cannot be explained by assuming that the many replicating units of DNA among the 18 chromosomes of the yeast nucleus all initiate replication at the beginning of the S period; purified DNA molecules from yeast contain "replication bubbles" of different sizes, indicating that yeast chromosomes contain multiple replicating units that initiate at different times (339).

RNA Synthesis

The continuation of DNA synthesis is also dependent upon RNA synthesis (see 444), but the dependency is far less immediate than it is in the case of protein synthesis. In HeLa cells, addition of actinomycin D early in S prevents the replication of DNA late in S (334). Addition of actinomycin D "two hours or more after initiation of DNA synthesis . . . has little effect on DNA synthesis."

Fujiwara (159) has also noted in L cells that later DNA synthesis is not inhibited by addition of actinomycin D in early S. Fujiwara also made an interesting observation with a level of actinomycin D (0.1 μg/ml) that might be expected to inhibit primarily ribosomal RNA synthesis, with much less effect on non-nucleolar RNA synthesis. He discovered that the low level of actinomycin D, given during the 1 to 3 hour interval after mitotic shake-off, prevented initiation of DNA synthesis in 50% of the cells; inhibition apparently occurred only in the cells that were the farthest back from the G_1–S border at the time of drug treatment. Exposure to 0.1 μg/ml of actinomycin D over the period of the G_1 to S transition had no effect on initiation of DNA synthesis. Rickinson (407) has followed up this study, again with a level of actinomycin D that "causes an apparently 'nucleolar-specific' inhibition of RNA synthesis" in L cells. His results disagree with Fujiwara's since, at this drug level, entrance into DNA synthesis was blocked or seriously delayed, but once DNA synthesis had been initiated, the low drug level did not apparently affect the rate of progress through S.

Thus, it would appear that ribosomal RNA synthesis must continue to within an hour or two of the G_1 to S transition if DNA synthesis is to be initiated. The last 1 to 2 hours of G_1 and the S period itself apparently may progress at a normal rate without ribosomal RNA synthesis. As already noted, at higher levels of actinomycin D the S period cannot be completed, perhaps because of the loss of synthesis of messenger RNA (mRNA) (for example, synthesis of mRNA for histones or inducers of DNA replication). All these studies do provide orientation for the design of perhaps more incisive analyses of the role of ribosomal and nonribosomal RNA synthesis in the initiation and continuation of the S period.

THE LENGTH OF THE S PERIOD IN CELLS OF DIFFERENT PLOIDIES

A further clue about the regulation of DNA synthesis is provided by measurements of the length of the S period in cells of different ploidies. Troy and Wimber (510) demonstrated for a series of different plants that the S period was the same length in both the diploid and autotetraploid versions of a given species. In *Xenopus*, haploid cells have half the volume and half the mass of their diploid counterparts, yet the S period occupies the same position in the cell cycle and lasts for the same period in both (180). In *Avena* (oat), diploid and autotetraploid cells have the same S period (546), and the same is true for diploid and tetraploid cells of bean roots (155). These results, as well as the evidence that homologous parts of homologous chromosomes replicate in synchrony (except the two X chromosomes in an XX cell), fit with the concept of banks or families of replicons in which the members of a bank initiate replica-

TABLE I

Comparison between the Mean Durations of Various Phases of the Cell Cycle (in Hours) and Nuclear DNA Amount[a]

	DNA/nucleus (pg)	G_2	D	G_1	S	Cell cycle time
Newt	45.0	0.9	3.5	30.0	41.0	79.0
Frog	14.6	1.15	3.1	19.2	26.0	52.0
Lizard	3.2	0.9	1.6	8.5	15.0	28.0
Chicken	2.5	1.05	0.5	3.4	6.9	12.5

[a]From Grosset and Odartchenko (191).

tion in synchrony. According to this concept, changing the size of a bank by decreasing or increasing the ploidy of a cell would not be expected to change the length of the S period.

The lack of a change in the length of S in relation to ploidy is in contrast to a putative relationship between length of S and genome size. Comparing four species of vertebrates (newt, frog, lizard, and chicken) with widely different genome sizes the length of S of erythroblasts increased with increasing amounts of DNA in the genome (Table I) (191). Increases in S period durations could be due to slower rates of replication fork travel, increased lengths of replicons, or larger numbers of replicons that are initiated over a longer time span, etc.

7

The G_2 Period

PREPARATION FOR MITOSIS

In the simplest view, the G_2 period represents the time required for a cell to synthesize elements necessary for chromosome condensation and the construction and operation of the mitotic apparatus. In some kinds of cells (perhaps only those that lack a G_1 period), the G_2 period is more complex since it probably contains events that ordinarily occur in the G_1 period in other cells (see earlier discussion of G_1-less cell cycles, Chapter 4). In a few situations the end of the S period leads immediately into prophase, and hence the G_2 period is altogether absent (164, 284). These observations lead to the speculation that the G_2 period in most cell types is concerned with initial stages of condensation of the chromosomes not visible in the light microscope, and hence G_2 is no more than an early part of prophase. Whether or not this is so is a less important matter than the actual sequence of molecular events by which the nucleus moves from the end of the S phase into the division configuration.

It is ordinarily assumed that the mitotic condensation of the chromosomes is brought about by the synthesis or activation of one or more factors at the point of transition of the cell into prophase. Johnson and Rao (247) have extended our knowledge of this situation by showing that the mitotic cell (HeLa) has the capacity to induce the condensation of the chromosomes of any interphase nucleus that is introduced into the mitotic cell by cell fusion [see also the follow-up study of Matsui *et al.* (305)]. For example, chromosomes in a G_1 nucleus of a HeLa cell condense into the mitotic form when the G_1 cell is fused to a metaphase cell. The condensation in such cases apparently follows the same progressive course that occurs in normal prophase. Each of the condensed G_1

chromosomes consists of only one chromatid, so obviously the chromosomes need not have been replicated in order to respond to the condensation factor(s) in the metaphase cell. The condensation factor(s) in HeLa metaphase cells is remarkably effective across distant phylogenetic lines. The chromosomes in bull sperm, chicken erythrocytes, cells of *Xenopus,* and mosquito cells were all induced to undergo mitotic condensation under the influence of the metaphase HeLa cell (248). We know nothing of how the condensation factor is caused to appear in prophase or how the disappearance or inactivation of the condensation factor is accomplished in the normal completion of mitosis.

ARREST OF THE CYCLE IN G$_2$

Whatever the program of events in G$_2$ it can be enormously slowed or blocked completely in some types of plant (467) and animal (264) cells for reasons that are not understood. The best studied case of the G$_2$ arrest or delay occurs in the ear epidermis in the mouse, the tissue in which the phenomenon was discovered by Gelfant (170). Typically, a few percent of the cells in the renewing epithelium are arrested in G$_2$, and these are the first cells to enter mitosis when the epithelium is stimulated by wounding. According to Pederson and Gelfant (370) epithelial cells may remain blocked in G$_2$ indefinitely (if the tissue remains undisturbed). In animals the G$_2$ arrest ordinarily involves only a few percent of the cells in a population and probably plays only a minor role in the regulation of the rate of cell renewal in certain tissues. Increasing the environmental temperature from 22 to 35°C for intact mice causes the release of G$_2$ blocked cells in ear epidermis (169). G$_2$ arrest may be a more common phenomenon in some plant tissues (515, 523). Experiments on pea seedlings indicate that cotyledons of pea seedlings produce a substance that is transported to the root where it promotes arrest of meristematic cells in the G$_2$ period (134). Whatever the significance of the G$_2$ arrest, the phenomenon demonstrates that the G$_2$ period contains a point at which the cycle may be interrupted without detrimental effects on cell viability. This reversible arrest of cells in G$_2$ is possibly mediated by cyclic AMP (348, 540).

A type of G$_2$ arrest, possibly related to the phenomenon just discussed, occurs when a G$_2$ cell is fused with a cell from an earlier part of the cycle. For example, in *Physarum* when a G$_2$ plasmodium is fused with a plasmodium in an earlier cycle stage, the G$_2$ nuclei are delayed in their entry into mitosis until the nuclei of the earlier plasmodium catch up (423). The same G$_2$ delay and synchronization at mitosis has been described for HeLa cell fusions (402). The cell at an earlier stage of the cycle is obviously able to block the G$_2$ cell from entry in mitosis (401). The molecular basis for the block is unknown, but it may be suggested that the activity of one or more of the cell cycle genes responsible for G_2 traverse is repressed by a factor(s) present in cells in G_1 and S.

REQUIREMENTS FOR PROTEIN AND RNA SYNTHESES DURING G_2

Taylor (486) used puromycin and chloramphenicol to show that protein synthesis is necessary to complete most of the G_2 period, but that "any protein synthesis necessary for mitosis is completed before the beginning of prophase." Sisken and Wilkes (456) and Sisken and Iwasaki (454) concluded from studies with amino acid analogues that protein essential for normal mitosis is synthesized in G_2 and perhaps also in the early part of prophase. According to their data, these particular division-related proteins are not synthesized *prior* to the G_2 period. This agrees with the discovery of Schindler *et al.* (434) that the arrest of cells in S with amethopterin for as much as 8 hours does not result in a shortening in the G_2 period when the cells are subsequently released from the block. The authors interpret this as evidence that the G_2 functions that establish the length of G_2 cannot begin in S, even when the cells are held for many extra hours in the late S period. These experiments of Schindler *et al.* (434) therefore suggest that the completion of DNA synthesis is the trigger for the initiation of the events specific to the G_2 period.

The earlier work of Taylor (486) and Kishimoto and Lieberman (259) on the blockage of transition from G_2 to mitosis by inhibitors of protein synthesis has been extended in detail by Tobey *et al.* (502) on Chinese hamster cells. They obtained evidence that protein synthesis is necessary until 10 minutes before the G_2-prophase border in order for cells to enter mitosis. However, for mitosis to go to completion, protein synthesis is necessary up until 2 minutes before the G_2-prophase border.

A number of studies (22, 122, 259) have shown that RNA synthesis is also necessary to complete G_2, but the requirement for RNA synthesis is fulfilled well in advance of the requirement for protein synthesis (502). According to the latter authors, in CHO cells RNA synthesis necessary for division is completed 1.87 hours before metaphase, while protein synthesis is required until 1.1 hours before metaphase. Donnelly and Sisken observed that 0.04 μg/ml of actinomycin D (a drug concentration that selectively inhibits the synthesis of ribosomal RNA) did not affect the entry into mitosis of cells that were within about 3 hours of reaching mitosis, while 4 μg/ml of actinomycin D (inhibition of all RNA synthesis including mRNA) stopped all cells that were not in the last 30 to 40 minutes of prophase. Similarly, in L cells traverse through G_2 and entry into mitosis occurs at the normal rate when ribosomal (nucleolar) RNA is selectively inhibited with 0.04 μg/ml of actinomycin D (407). Inhibition of ribosomal RNA synthesis in S causes a delay in advancement of cells to mitosis. In plants, inhibition of RNA synthesis with 3'-deoxyadenosine has indicated that RNA synthesis must continue up to early prophase in order for cells to reach metaphase (174).

All these results, most of which are reviewed by Tobey *et al.* (500), indicate that transcription of RNA (mRNA?) is required at least into mid to late G_2, and protein synthesis is required until very late G_2. Kolodny and Gross (266) have made a step toward characterization of proteins that are synthesized in G_2. They have shown by electrophoretic analysis that there are at least several proteins that are synthesized primarily or exclusively in the G_2 period of HeLa cells. The function of the G_2-specific proteins is not known, but there are some obvious possibilities, for example, proteins for construction and operation of the mitotic apparatus or proteins needed for chromosome condensation.

THE G_2 TO D TRANSITION

Theoretically, at least, the G_2 period ends with the beginning of the mitotic condensation of the chromosomes, but the position of the transition cannot be identified very precisely because condensation must begin some minutes before it can be identified by microscopy. Again, it is usually assumed that the initiation of chromosome condensation is a visible manifestation of the readout of a program of gene transcriptions that carries the cell step by step through the cycle, although there is no direct proof. Considerable evidence points to the phosphorylation of F_1 histone as a key event in the transition from G_2 into mitosis (see Chapter 11).

8

Activities during Cell Division

Large and rapid changes in cell structure and function occur during mitosis, but we still understand relatively little of the molecular basis or the regulation of these changes.

RNA SYNTHESIS DURING MITOSIS

The rate of RNA synthesis declines rapidly in late prophase and all RNA synthesis stops before metaphase is reached (Fig. 43) [see review by Prescott (392)], with the exception of the continued synthesis of some 4 S and 5 S RNA during mitosis (560). RNA synthesis resumes in late telophase and rapidly rises to a more or less steady rate within minutes after telophase is over. The mechanism by which nuclear RNA synthesis is blocked during mitosis is not known. The block has been ascribed to the unavailability of DNA templates for RNA transcription when the chromosome is in the tightly condensed mitotic form, and this is true at least in part (145). When chromatin is reconstituted from pure DNA and proteins derived from mitotic chromosomes of HeLa cells, the template activity is much reduced compared to chromatin reconstituted with proteins from S period cells. Thus, proteins of mitotic chromosomes are effective repressors of transcription in general, although these proteins do not completely block the synthesis of 4 S and 5 S RNA on mitotic chromosomes.

At about the same time that the synthesis of most classes of RNA stops in late prophase, much of the protein of the nucleus is released into the cytoplasm. Some of the released proteins might possibly be essential for RNA synthesis. Possibly the association with DNA of RNA polymerase and other putative

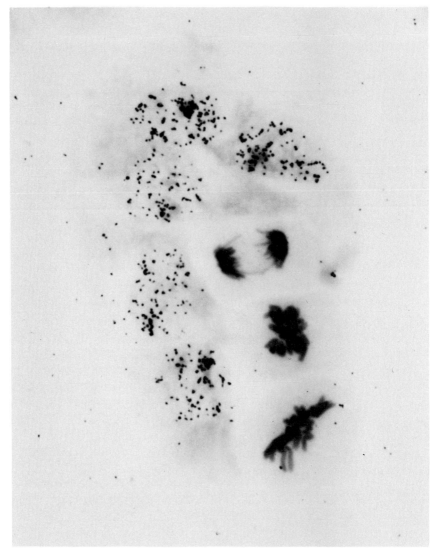

Fig. 43. Cessation of RNA synthesis during mitosis. Chinese hamster ovary cells were labeled for 15 minutes with ^{3}H-uridine and then autoradiographed. Cells in mitosis fail to incorporate radioactivity. The amount of synthesis of 4 S and 5 S RNA reported by Zylber and Penman (560) for mitotic cells is apparently too small to be detected in this autoradiographic study.

factors involved in transcription is physically incompatible with an advanced state of chromosome condensation.

PROTEIN SYNTHESIS DURING MITOSIS

About the time that most RNA synthesis stops in late prophase the rate of protein synthesis begins to fall, dropping by about 75% by late telophase (231, 394, 427, 475). Stein and Baserga (475) have presented evidence that the rate of synthesis of chromosomal proteins does not decline during mitosis. From their data it appears that almost half the protein synthesis during mitosis can be accounted for by these chromosomal proteins. In late telophase or perhaps shortly thereafter, the rate of total protein synthesis increases rapidly, reaching a more or less steady level in very early interphase.

The decline in protein synthesis during mitosis has been ascribed to the shutoff of mRNA synthesis during mitosis, but this is probably not a correct explanation. First, the half-life of mRNA in eukaryotic cells is far in excess of the time taken for mitosis. Second, mRNA in HeLa cells persists through mitosis, yet the polyribosomes disaggregate during mitosis (133, 231, 429, 481). This disappearance of polysomes is due to a decreased ability of ribosomes to become attached to mRNA, which is equivalent to an inhibition of the initiation of translation of mRNA during mitosis (142). Ribosomes purified from HeLa cells in metaphase are one-third as effective as ribosomes from interphase cells in supporting synthesis of polyphenylalanine on polyuridylic acid templates (427). Gentle treatment with trypsin restores to the interphase level the effectiveness of purified metaphase ribosomes for polyphenylalanine synthesis. Thus, metaphase ribosomes appear to be blocked by a protein, perhaps one of the proteins released from the nucleus during mitosis. The ribosomes become unblocked when the postmitotic nuclei are reformed, presumably because the blocking protein, like all the released nuclear proteins, returns to the reforming nucleus.

Polyribosomes reform and the rate of protein synthesis rises in postmitotic cells derived by allowing metaphase cells to complete mitosis in the presence of actinomycin D (481). When actinomycin D is added to HeLa cells in metaphase, the synthesis of some classes of nonhistone chromosomal proteins nevertheless takes place in the subsequent G_1 period (476), indicating the carry-over through mitosis of the corresponding mRNA's that must have been synthesized in the G_2 period. Third, Salb and Marcus (427) found that ribosomes prepared from HeLa cells in metaphase could support *in vitro* protein synthesis at only a fraction of the rate supported by ribosomes isolated from interphase cells. The competence of ribosomes from mitotic cells could be restored by gentle trypsin treatment. Thus, it must be concluded that the decreased rate of protein synthesis is not due to lack of mRNA but to a transient incompetence of the ribosomes.

Concomitant with these changes, the mitotic cell loses the ability to support the replication of vaccinia virus (193), presumably because the cellular ribosomes cannot be recruited by the virus for protein synthesis.

REVERSIBLE SHIFT OF NUCLEAR RNA AND PROTEIN TO THE CYTOPLASM DURING MITOSIS

Concurrent with the cessation of almost all RNA synthesis and the slowdown in protein synthesis, a variety of other changes occurs. The nuclear envelope becomes fragmented, the nucleolus begins to break down, and a large amount of nuclear material is released to the cytoplasm. In amoeba, more than 90% of the nuclear proteins become dispersed into the cytoplasm in late prophase (389). The remaining proteins are presumably all a part of the mitotic chromosomes. In various kinds of vertebrate cells it appears that at least half the nuclear proteins are shifted to the cytoplasm during mitosis (35). In plant endosperm cells 75% of the nuclear mass is lost to the cytoplasm (406); a major part of this shifting material must be protein. Massive migration of proteins during mitosis has also been demonstrated by Sims (453) in rat cells. In several of these studies there were indications that the proteins that had dispersed to the cytoplasm during mitosis returned to the reforming nucleus in late telophase and early interphase. This is clearly the case with amoeba; half the proteins return to the nucleus within the first hour of interphase, and all the proteins are back in the nucleus by 3 hours (Fig. 44). Also, the protein that is the nuclear T antigen of SV40 transformed cells becomes dispersed throughout the cytoplasm during mitosis and returns to the nucleus when mitosis is completed (173, 479). Presumably, the localization of nuclear proteins in the interphase nucleus is the result of high affinity of these proteins for the dispersed chromosomes of interphase.

It has long been known that most of the RNA is lost from the nucleus during mitosis [see Rao and Prescott (400), for a brief review]. Nucleolar material has often been described to spread out as a film on the surface of mitotic chromosomes, and this may account for the RNA that is associated with the chromosomes during division (see 388). This is borne out by the observation that ribosomal precursor RNA's are associated with the chromosomes isolated from metaphase mammalian cells (141). The processing of ribosomal precursor RNA's is arrested during mitosis (4). The fate of the bulk of nuclear RNA has not been determined in detail, but it is clear from autoradiography that it becomes dispersed into the cytoplasm. The high molecular weight heterogeneous nuclear RNA (HnRNA, which normally turns over rapidly within the interphase nucleus) can be recovered from the cytoplasm of mitotic mammalian cells. HnRNA is apparently not degraded while it resides in the cytoplasm since it persists

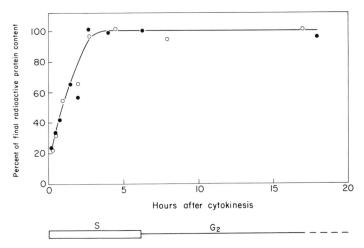

Fig. 44. Return of labeled proteins to the nucleus of amoeba after mitosis. At least 90% of the nuclear proteins leave the nucleus during late prophase, and all of these return to the postmitotic daughter nuclei. Some protein has returned to daughter nuclei by the time of the first measurements in this experiment. Redrawn from Prescott and Goldstein (389).

quantitatively for at least 3 hours in metaphase cells arrested with colchicine (D.M. Prescott and M.R. Lauth, unpublished). The rapid turnover of HnRNA and the processing of ribosomal precursor RNA's appear to require the intact nucleus of interphase. Similar results have been obtained on Chinese hamster cells by Abramova and Neyfakh (4).

In amoeba virtually all nuclear RNA is released to the cytoplasm in late prophase (400). Within 15 minutes after cytokinesis (late telophase–early interphase) 35% of the released RNA returns to the two reforming daughter nuclei (17.5% to each). No further RNA moves into the nucleus after this initial rapid return. In fish embryos (343) and Chinese hamster fibroblasts (342), most, if not all, of the RNA synthesized shortly before mitosis (as opposed to total nuclear RNA in the case of amoebae) and released to cytoplasm during mitosis rapidly returns to the reforming daughter nuclei in late telophase. The return of ribosomal precursor RNA to the postmitotic nucleus probably contributes to the reconstitution of the nucleolus, accounting perhaps for the observation that at least partial nucleolar reconstitution takes place in the absence of postmitotic RNA synthesis (377, 378, 480). Once back in the postmitotic nucleus the processing (turnover) of the RNA is apparently resumed (4, 342).

Thus, it appears that during late prophase a large proportion of nuclear RNA and protein is released to the cytoplasm, and beginning about the same time the synthesis activities of the cell are drastically reduced. As a part of the reformation of the daughter nuclei in late telophase, nuclear protein and RNA in the cytoplasm return to the nucleus, and rates of synthesis of protein and RNA rise sharply, i.e., the events of late prophase occur in reverse in late telophase. The molecular mechanisms that underlie the program of mitotic events and their reversal in late telophase are completely unknown.

9

Cell Surface Changes during the Cycle

MORPHOLOGICAL CHANGES

The cell surface undergoes morphological changes during the cycle that are temporally related to intracellular events (Fig. 45) (3, 204, 384) although the causal connection between the surface changes and intracellular activities is not known. In Chinese hamster cells the transition from G_1 to S is characterized by the disappearance of surface projections (various blebs, villae, and filapodia) so that the surface of the S phase cell is almost smooth and featureless. During G_2, microvilli increase in number, and the cells thicken as they gradually approach the "rounded up" stage of mitosis. In late G_2 long filapodia appear, and the surface of cells in mitosis has an abundance of these structures, some of which hold the cell to the substrate. This succession of changes in surface morphology takes place in relation to a succession of events within the cell, but the causal connections between the two are completely unknown. A main difficulty in understanding the surface change in terms of the cycle is that the functional significance of the microvilli, blebs, filapodia, and ruffles is not known. The observations nevertheless affirm that the cell surface is in some way related to the progression of changes in activities occurring within the cell as it moves through its cycle. These morphological changes during the cell cycle are also related to the density of cells in the culture. In very sparse cultures of the same cell line, the cells remain in the less flattened blebbed form of G_1 throughout the cycle (417).

The cell cycle changes in surface morphology observed by Porter et al. (384) indicate that if the blebs and villi are taken into consideration the total surface

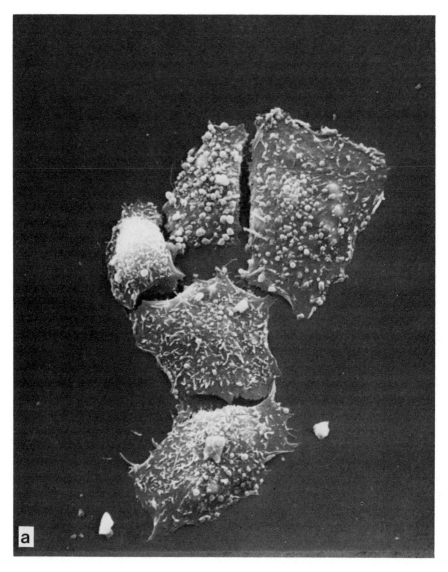

Fig. 45. Changes in the surface morphology of Chinese hamster cells during the cell cycle observed by scanning electron microscopy. (a) A group of cells in mid-G_1 (3 hours after mitotic selection). From Porter *et al.* (384).

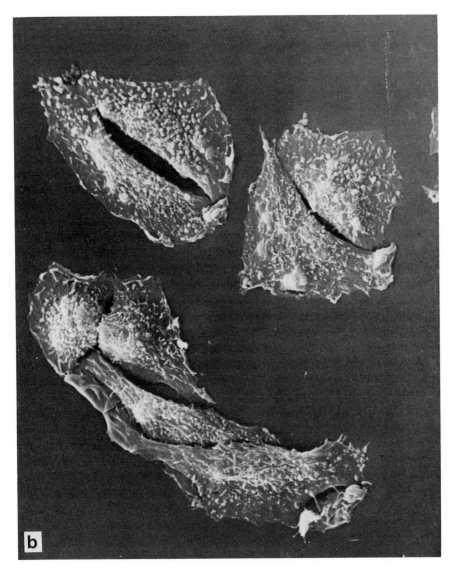

Fig. 45. (b) A group of cells in late-G_1 (5 hours after mitotic selection). From Porter *et al.* (384).

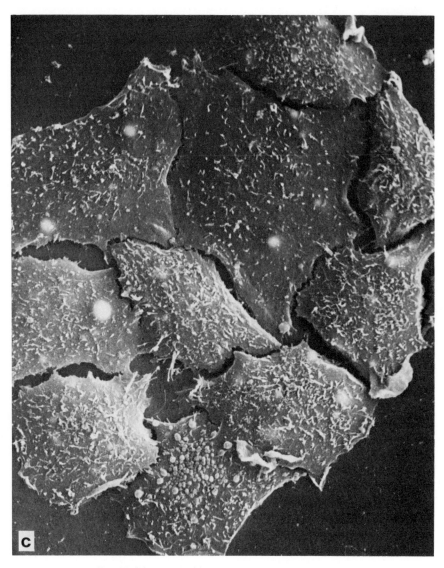

Fig. 45. (c) Cells in the S period. From Porter *et al.* (384).

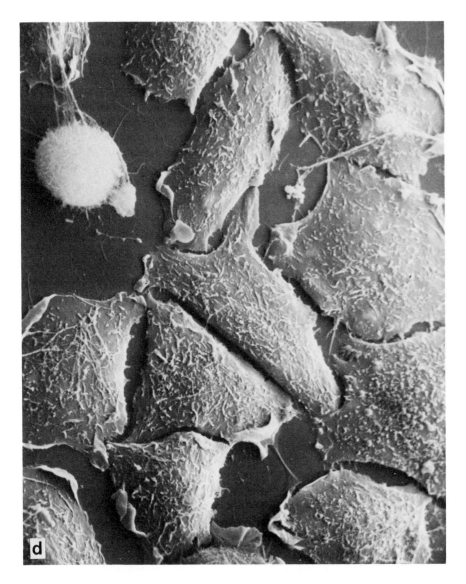

Fig. 45. (d) Cells predominantly in G_2 period; a cell in the upper left has entered prophase. From Porter *et al.* (384).

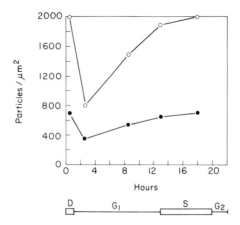

Fig. 46. The population density of intramembranous particles in the plasma membranes of L cells synchronized by amethopterin inhibition of TMP synthesis (arrested near the G_1–S border). Open circle, inner fracture face; closed circle, outer fracture face. Redrawn from Scott *et al.* (441).

area is probably considerably greater in G_1 than in S phase cells, even though S phase cells have a larger volume. The total surface area again increases in G_2.

It would be useful to know whether distinct changes in surface morphology occur when the same cells are grown in suspension. The results of Rubin and Everhart (417) suggest that cell contact is necessary for the changes to occur,

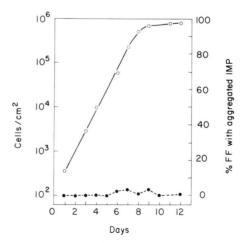

Fig. 47. Growth curve for SV3T3 cells (open circles). In contrast to normal 3T3 cells, the intramembraneous particles (IMP) of the plasma membrane do not aggregate as cell population density increases and cells come into contact (closed circles). FF, fracture faces of plasma membranes. Redrawn from Scott *et al.* (440).

and hence cells growing in suspension may behave as cells in sparse surface cultures.

Chicken embryo cells (normal) show cell cycle changes similar to those of Chinese hamster cells (transformed) (204). Early G_1 cells have numerous blebs, S cells are relatively smooth, and G_2 cells are covered with many microvilli.

Another kind of morphological change in the plasma membrane during the cell cycle has been described for intramembranous particles, observed by freeze-fracture, of mouse L cells grown in suspension (441). From late G_1, through S, and into metaphase, the pattern of intramembranous particles is dense (Fig. 46). In the transition from mitosis to G_1 the density of the particle population drops two- to threefold, to increase again through G_1.

The pattern of intramembranous particles also changes in relation to cell contact of 3T3 cells growing in monolayers (440). In noncontacted 3T3 cells the intramembranous particles of the plasma membrane are randomly distributed in the plasma membrane. As the cells proliferate and begin to establish contacts, the percentage of fracture faces of plasma membranes with aggregated particles increases. The aggregation of intramembranous particles precedes the density-dependent inhibition of cell reproduction (G_1 arrest) which could be taken to mean that aggregation is not related to regulation of cell reproduction. However, 3T3 cells transformed by SV40, cells that do not show density-dependent inhibition of the cell cycle, also fail to show aggregation of intramembranous particles even at high cell densities (Fig. 47).

In summary, the morphology of the cell surface and the structure of the plasma membrane undergo regular changes in relation to the cell cycle, but the meaning of these changes will probably remain obscure until the causal links between cell metabolism and the cell surface can be defined.

CHEMICAL CHANGES OF THE PLASMA MEMBRANE

The amount of protein and phospholipid in the plasma membranes of cells grown in suspension (transformed hamster fibroblasts–NIL-2HSV) remains proportional to total cell protein during the cycle and hence increases gradually and doubles between early G_1 and late G_2 (Fig. 48) (184). Plasma membrane carbohydrates (mouse mastocytoma cells–P815Y, grown in suspension) increase at a faster rate in G_1 and in S; the rate again rises in G_2 (Fig. 49).

In HeLa cells arginine moieties in protein at the cell surface decrease sharply as cells go from mitosis to G_1 (477). When the cells enter S the detectable arginines increase twenty-four-fold and remain high until the next mitosis. In contrast, tyrosine moieties of surface proteins increase gradually with increase in cell size. These facts indicate that arginine-rich proteins are present on the cell surface during S, G_2, and mitosis but are withdrawn from the surface or are somehow

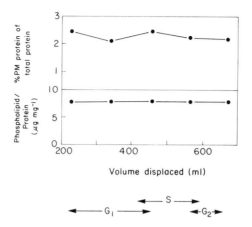

Fig. 48. Amount of protein in the plasma membrane as a percentage of total cell protein and amount of plasma membrane phospholipid phosphorus in micrograms per milligram of total cell protein in relation to the cell cycle in transformed hamster fibroblasts (NIL-2HSV). The measurements were made on cells grown in suspension and separated into cell cycle groups by size by velocity sedimentation. Redrawn from Graham *et al.* (184).

masked during G_1. The significance of this change is not known but it provides a clear example of a change in membrane proteins in relation to the cell cycle.

PLANT LECTIN BINDING AND CELL AGGLUTINABILITY

In normal cells, membrane sites for binding of plant lectins are transiently present in a higher concentration during mitosis (82, 149, 347). The experiment

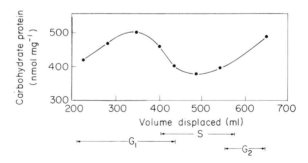

Fig. 49. Amount of carbohydrate in the plasma membrane per total cell protein (nmoles/mg) in mouse mastocytoma cells (P815Y) separated into cell cycle groups by size by velocity sedimentation. Redrawn from Graham *et al.* (184).

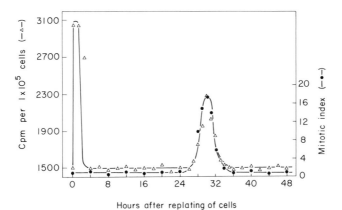

Hours after replating of cells

Fig. 50. Increased binding of ³H-concanavalin A to the surface of 3T3 cells during mitosis. The increased binding at zero time is due to the trypsin treatment used in replating. Redrawn from Noonan *et al.* (347).

in Fig. 50 showing ³H-concanavalin A binding in mouse 3T3 cells is from the study of Noonan *et al.* (347). Cells were grown to confluency, which results in G₁ arrest of 3T3 cells. The cells were released from G₁ by replating at lower density. A burst of mitosis occurred between 28 and 32 hours after replating. The binding of ³H-concanavalin A increased transiently at replating because of the brief trypsin treatment used to remove the cell monolayer. The binding of ³H-concanavalin A then increased transiently in parallel with the burst of mitosis.

During the cell cycle the agglutinability of cells by plant lectins such as concanavalin A parallels the ability of cells to bind the lectin. Normal cells have low susceptibility to agglutination during interphase and become agglutinable during mitosis. Transformed cells retain the same high agglutinability throughout the entire cell cycle (see 82, 422).

Smets and DeLey (459) and Smets (460) have carefully measured the agglutinability of cells with concanavalin A over the cell cycle and concluded that the main difference between normal and transformed cells is in the G₁ period. Thus, normal cells quickly lose agglutinability after mitosis, but transformed cells remain agglutinable during G₁ and slowly lose agglutinability through S and G₂ (Fig. 51).

The arguments about how much change in agglutinability of normal cells during mitosis is due to unmasking of agglutination sites on the surface versus rearrangement (aggregation) of sites are reviewed by Burger (82). The increase in cell agglutinability by plant lectins that accompanies loss of cell growth regulation has been reviewed by Sachs (422).

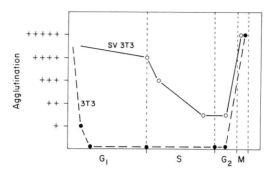

Fig. 51. Agglutination with concanavalin A during the cell cycle of SV3T3 and 3T3 cells. Redrawn from Smets and DeLey (459).

In any case, the transient change in surface chemistry during mitosis appears somehow related to regulation of the cell cycle, since a brief treatment with trypsin of normal cells arrested by contact inhibition in G_1 results in the transient increase in concanavalin A binding (Fig. 50), the appearance of agglutinability, and the release from G_1 arrest. Perhaps related to the transient state of agglutinability during mitosis is the discovery of Kraemer and Tobey (268) that surface heparan sulfate is released into the medium during mitosis in Chinese hamster cells. Whether heparan sulfate is similarly released by normal cells, however, is not known.

10

Cyclic AMP, Cyclic GMP, and Cell Reproduction

A variety of studies points strongly to cyclic AMP (cAMP) as an important factor in the regulation of cell reproduction (see reviews 2, 82). More limited information suggests that cyclic GMP (cGMP) may also be involved in growth regulation, apparently affecting cell metabolism in an opposite or antagonistic direction to cAMP.

cAMP AND THE CELL CYCLE

The largest change in the intracellular concentration of cAMP observed consistently is a pronounced decrease during mitosis. This occurs in mouse 3T3 cells and in Py3T3 cells (transformed with polyoma virus) synchronized by several methods (Fig. 52) (83). A similar drop in cAMP concentration occurs in Chinese hamster ovary cells (spontaneously transformed) during mitosis (Table II) (449). In this study, the level of cAMP increased threefold immediately after mitosis. Between the first hours of G_1 and late G_1, the concentration dropped to an intermediate level that remained about the same when the cells entered the S period. In general, higher intracellular concentrations of cAMP occur in cells that are blocked in the cell cycle for one reason or another. It may be that the transient high concentration of cAMP in early G_1 is involved in the putative transient block in the cell cycle postulated as responsible for the high variability of the G_1 period (Chapter 4).

Similarly, Millis et al. (315) detected a G_1 rise in cAMP in a lymphoid cell line but in this study did not attempt to show whether the rise was associated with a particular part of G_1. They also observed a higher concentration of cAMP in the

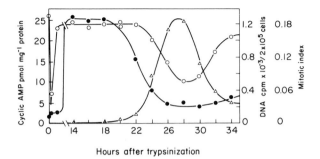

Hours after trypsinization

Fig. 52. Changes in the intracellular concentration of cAMP in mouse 3T3 cells. Open circles, cAMP; closed circles, DNA synthesis; open triangles, mitotic index. The cells were synchronized by trypsinization of a confluent monolayer (arrested in G_1) and replating in fresh medium at a 5-times-lower cell density. Redrawn from Burger *et al.* (83).

G_2 period as compared with the S period or mitosis. This leads to the speculation that cAMP might have something to do with the G_2 arrest sometimes observed for a small fraction of cells in various tissues and cultured cells (370). Ordinarily, however, it appears that cells are able to pass the arrest point in G_2 to become stopped at a putative cAMP-sensitive point early in the subsequent G_1 period.

In contrast to the work on the Chinese hamster ovary cell line, measurements on the V79 line of Chinese hamster cells failed to show a rise in cAMP during early G_1 (421). Instead the concentration rose steadily through G_1, reached a peak in early S, and declined through the rest of interphase. This is somewhat similar to a pattern observed in HeLa cells (550). Relative to a value of 1.0 during mitosis, the cAMP level rose to 1.6 in early G_1, reached a peak of 2.6 around the G_1–S border, fell to 1.4 in early S, rose slightly (1.8) in late S, and fell again to 1.2 in mid G_2.

Cell cycle fluctuations of cAMP in the cleavage stage of sea urchin embryos are extraordinary (Fig. 53) (547). Fifteen minutes after fertilization the concentra-

TABLE II
cAMP Concentration in Chinese Hamster Ovary Cells[a]

Phase of cell cycle	μM of cAMP	pmoles cAMP/mg protein
Mitosis	2.9 ± 0.4	16 ± 2
Early G_1	8.4 ± 0.9	44 ± 8
Late G_1	4.0 ± 0.5	24 ± 4
S	5.2 ± 0.4	28 ± 4

[a]From Sheppard and Prescott (449).

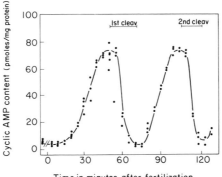

Time in minutes after fertilization

Fig. 53. Fluctuation in the content of cAMP in sea urchin embryos from fertilization through the second cleavage. Redrawn from Yasumasu *et al.* (547).

tion of cAMP begins to rise and increases fourteenfold by the beginning of cell division. During the first cell division the cAMP drops rapidly to the initial level. The pattern is repeated at the second and third cleavages. Addition of 10^{-3} M caffeine to fertilized eggs causes the level of cAMP to rise to an even higher level, presumably by blocking phosphodiesterase-catalyzed destruction of cAMP, abolishes the decrease observed in untreated eggs, and inhibits cell division.

The intracellular level of cAMP in a human lymphoid cell line released from a thymidine block has been compared with the enzymes that catalyze the synthesis and destruction of cAMP (314). The level of phosphodiesterase was high immediately after release of cells arrested near the G_1–S border and then fell to its lowest level at 2 and 3 hours after release (late S period) (Fig. 54). During the first 3 hours after release from the thymidine block, both cAMP and adenyl cyclase were low; both rose sharply in late S and in G_2 and finally fell rapidly as the cells entered mitosis. As cAMP and adenyl cyclase fell, the level of phosphodiesterase rose to its maximum. The observation of a low amount of cAMP in S period cells (compared with mitotic cells) is not consistent with patterns observed by others. However, the activities of adenyl cyclase and phosphodiesterase conform to what would be predicted on the basis of the high cAMP level in G_2 and the low level in mitosis. Contrary to this expectation the peak of adenyl cyclase activity in synchronized liver cells in culture has been reported to be present during mitosis (300).

Information obtained so far indicates that the concentration of cGMP also changes during the cell cycle, but in a pattern different from cAMP. In synchronous 3T3 cells, cGMP increased transiently by about tenfold in mid G_1; through the rest of the cycle, including mitosis, the concentration of cGMP remained more or less constant (Fig. 55) (442). In the same experiment, cAMP rose from a low level in mitosis to a threefold higher level in early G_1. During

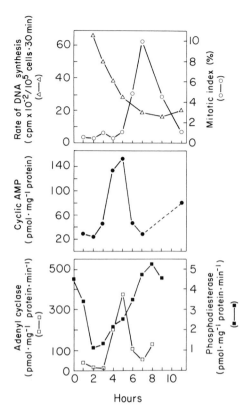

Fig. 54. cAMP (middle frame), adenyl cyclase (lower frame, open squares), and phosphodiesterase (lower frame, closed squares) in relation to the cell cycle (upper frame) of lymphoid cells. The cells were released from thymidine block at zero time. Redrawn from Millis *et al.* (314).

the transient rise in cGMP during mid G_1, cAMP decreased transiently. Seifert and Rudland (442) suggest that the transient increase in cGMP may "act as a specific signal for movement of cells out of the G_0 or G_1 phase of the cell cycle." This idea was derived, not only from the data in Fig. 55, but from the observations (a) that addition of cGMP or dibutyryl cGMP to confluent 3T3 cells (G_1 arrested) induces a large increase in DNA synthesis measured 12 to 30 hours later, and (b) that release of arrest of confluent cells by addition of fresh serum is quickly followed by an immediate dramatic rise in intracellular cGMP (Fig. 56) (443). During the same interval, the level of cAMP is transiently decreased. These changes precede the initiation of DNA synthesis by many hours, indicating that the arrest point controlled by the cyclic nucleotides is located some hours before the start of the S period.

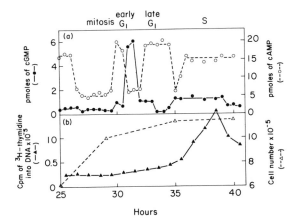

Fig. 55. (a) Cyclic nucleotides. Changes in cAMP (open circles) and cGMP (closed circles) in the second synchronous cell cycle after release of a confluent monolayer of 3T3 cells by addition of serum and fresh medium. (b) Cell growth characteristics. The increase in cell number (open triangles) represents the mitoses of the end of the first cell cycle after G_1 release. The curve for DNA synthesis (closed triangles) defines the second S period after G_1 release. Redrawn from Seifert and Rudland (442).

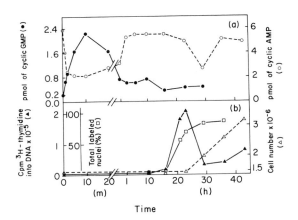

Fig. 56. Changes in cGMP (closed circles) and cAMP (open circles) in 3T3 cells released from density-dependent inhibition by addition of fresh serum. DNA synthesis (closed triangles); percent labeled nuclei (open squares); cell number (open triangles). Redrawn from Seifert and Rudland (443).

In summary, the level of cAMP does fluctuate during the cell cycle. A drop has consistently been observed during division. The level of cAMP may also fluctuate during interphase, but the data are less consistent, at least in part, because none of the various synchrony methods for cultured cells is very good and because blocking cells with inhibitors to synchronize cells near the $G_1 - S$ border could cause perturbations in cAMP metabolism for some period following reversal of the inhibition.

Transient high levels of cAMP in short portions of G_1 and G_2 might be associated with the ability of the cycle to be blocked at those points. The cell cycle data on cAMP and cGMP levels in themselves only suggest this possibility. Experiments on cAMP and cell growth discussed in the next two sections provide some additional support for this idea.

EFFECTS OF EXPERIMENTALLY INDUCED
INCREASES IN INTRACELLULAR cAMP ON CELL
GROWTH

Addition of cAMP (usually as dibutyryl cAMP because it penetrates the plasma membrane of most cells more readily than cAMP) to transformed and normal cells can inhibit their growth altogether or reduce the saturation densities of the transformed cells to those of their normal counterparts. Added to the earlier demonstrations of this effect [reviewed by Sheppard (447) and Abell and Monahan (2)] are reports of growth inhibition of mouse mammary carcinoma cultures (410), mouse L cells (352), Py3T3 cells (190), and cultures of human liver cells (125). In some of these experiments, theophylline, which inhibits phosphodiesterase, was added with dibutyryl-cAMP or cAMP. In mouse mammary carcinoma, theophylline alone was sufficient to inhibit growth (410). When growth of human liver cells was inhibited by labeled cAMP (used instead of dibutyryl-cAMP), none of the labeled cAMP could be found in the cells (125), suggesting that cAMP exerts its effect on growth by interaction with the cell surface. However, by fusion to cells of cAMP-containing phospholipid vesicles, it has been possible to circumvent the plasma membrane and introduce cAMP, as opposed to dibutyryl-cAMP, into mouse 3T3 cells and SV40 transformed 3T3 cells (366). Growth of both the normal and transformed cells was reduced by cAMP concentrations in vesicles of 10^{-7} M. A comparable effect on growth was achieved with dibutyryl-cAMP in the medium at a thousandfold concentration. These experiments appear to show that cAMP must enter cells to affect their growth.

Efforts to determine where in the cell cycle cAMP exerts its growth inhibitory effect have not yielded unambiguous results. L cells appear to be greatly slowed in S or blocked in G_2 (490). In cultured hepatoma cells, dibutyryl-cAMP appears to inhibit DNA synthesis (516). However, cAMP or dibutyryl-cAMP inhibits growth of human diploid fibroblasts (158), mouse lymphoma cells (104), and KB cells (491) by blocking in G_1.

cAMP does not block growth of all cell lines, at least not immediately. At least one line of Chinese hamster ovary cells continues to grow at the normal rate for two cell doublings or more after addition of dibutyryl-cAMP. Although growth was unaffected in this cell line, the cells immediately underwent a distinct morphological change (elongation into spindle-shaped cells and disappearance of surface blebs) in response to the dibutyryl-cAMP described in detail by Porter *et al.* (385). In these cells thymidine uptake was reduced twenty-one-fold by dibutyryl-cAMP at 1°C, suggesting an alteration in membrane permeability. In addition thymidine kinase activity was reduced in the presence of the cyclic nucleotide, and the pool size of thymidylate was strongly reduced. Thus, dibutyryl-cAMP can cause profound changes in this transformed cell without altering the cell cycle. These experiments also provide a warning about using incorporation of exogenously supplied ^{3}H-thymidine as a measure of cAMP effects on the rate or amount of DNA synthesis.

Similarly, the growth of monkey cells (line CV-1) appears unaffected by exogenous cAMP (411), although in contrast to Chinese hamster ovary cells the uptake of thymidine is stimulated by cAMP.

As described above, dibutyryl-cAMP arrests mouse lymphoma cells in G_1, although a mutant of this cell lacking cAMP-dependent protein kinase is not affected even by high levels of exogenous dibutyryl-cAMP (104). Loss of cAMP-dependent protein kinase activity could represent one of several kinds of metabolic lesions that could cause loss of cAMP-mediated regulation of the cell cycle (G_1 arrest) in transformed cells.

CYCLIC NUCLEOTIDES AND THE IMPOSITION AND RELEASE OF G_1 ARREST

Changes in intracellular concentrations of cAMP and cGMP induced by release of G_1-arrested cells were mentioned in the previous section. Since it has been generally assumed that the high level of cAMP in density-inhibited monolayer cells is responsible for the arrest of the cell cycle, proliferating cells should be expected to have a lower average content of cAMP. Several laboratories (327, 351, 448) have reported that cAMP levels in 3T3 cells do not change over a broad range of cell densities preceding or following arrest in a confluent monolayer. These results are contrary to those of Kram *et al.* (269), Rudland *et al.* (419), and Otten *et al.* (358), which indicated a higher level of cAMP in arrested cells compared to growing cells. In a repeat of Sheppard's (448) study, Bannai and Sheppard (30) found that the increase in cAMP occurs in response to cell contact but well before confluency is reached. Thus, they affirm that the increase "in cyclic AMP level has an important role in contact inhibition," as Otten et al. (358, 359) and Heidrick and Ryan (224) had concluded earlier. This is consistent with studies of adenyl cyclase, phosphodiesterase, and cAMP in normal rat kidney cells (NRK cells) (16). As NRK cells reach confluency, "phosphodiesterase activity decreases somewhat whereas adenyl cyclase activity continues to rise. This increase in synthetic ability is accompanied by the increase in cyclic AMP levels which occurs in these cells at confluency" (16). In chick embryo fibroblasts, which do not show density-dependent inhibition in confluent monolayers, both phosphodiesterase and adenylate cyclase activities increased proportionately with cell density, and there was no change in cAMP level. Even two studies that failed to detect an increase in cAMP in 3T3 cells as they became confluent (327, 351) did detect an increase in cAMP if growth was inhibited by serum restriction. Therefore, Oey et al. (351) conclude that "growth control in cultured fibroblasts is mediated through both density-specific and serum-specific regulations, and that sensitivity to serum-restriction, but not sensitivity to density-restriction, is reflected in a great rise in cAMP."

The release of density-dependent inhibition by addition of fresh serum or insulin or by trypsin treatment causes a rapid (within 5 minutes) but transient (a few hours at most) reduction in the level of cAMP (83, 269, 419, 443). The cAMP concentration returns to the initial level well before the first cells reach DNA synthesis. The release from G_1 arrest of confluent 3T3 cells with insulin or fresh serum can be blocked by addition of dibutyryl-cAMP to the medium (48).

Immediately following serum-induced reduction in cAMP in quiescent 3T3 cells, uridine and phosphate transport into the cell increase sharply (244, 415). The increased transport of uridine, but not of phosphate, is inhibited by stimulation of adenyl cyclase (by addition of prostaglandin E_1) or inhibition of phosphodiesterase (by addition of theophylline). These transport changes could be part of the chain of events that reinitiates the cell cycle, or they may be a secondary result of release from cell cycle arrest. Control of cell growth by regulation of the rate of nutrient uptake seems an unlikely hypothesis since in at least one case (223) dibutyryl-cAMP added to the medium reduces nutrient transport into the cell without reducing cell reproduction. It is likely that inward transport of nutrients by the plasma membrane is not limiting for cell metabolism but is adjusted to meet the particular metabolic needs of the cell. Evidence for this arrangement in the case of glucose has been assembled and reviewed by Elbrink and Bihler (126).

As already noted in the previous section (442, 443), cGMP increases transiently, and cAMP decreases within minutes after release of density-dependent inhibition (see also 327, 419). In lymphocytes stimulated to proliferate by phytohemagglutinin, the intracellular concentration of cGMP begins to rise immediately and increases seventeenfold within 20 minutes (Fig. 57) (202). The

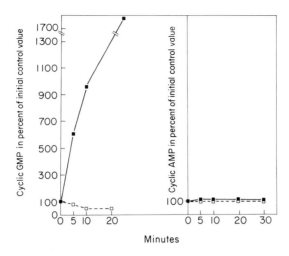

Fig. 57. Effect of phytohemagglutinin on levels of cGMP and cAMP in human lymphocytes. Closed symbols, cells treated with phytohemagglutinin; open symbols, control cells.
Redrawn from Hadden *et al.* (202).

level of cAMP, however, does not change during the first 30 minutes after
addition of phytohemagglutinin, supporting the suggestion that it may be the
cGMP/cAMP ratio that is crucial in release from G$_1$ arrest and reversal of the G$_0$
state rather than the absolute amount of either cyclic nucleotide. Since phytohemagglutinin stimulates lymphocytes to proliferate without entering the cell
(188), the rapid increase in cGMP must be triggered by an event in the cell
membrane (202). Furthermore, exogenously provided cGMP has also been found
to stimulate DNA synthesis in splenic lymphocytes, presumably by raising the
intracellular level of cGMP (525).

In a study on 3T3 cells already mentioned above, Rudland *et al.* (419) found
that not only did the cAMP rise as growing cells ceased to divide, but the level of
cGMP fell. In SV40-transformed 3T3 cells or polyoma-transformed BHK cells,
which appear to arrest at random in the cell cycle when growth conditions
become limiting, changes in cAMP and cGMP did not occur as growing cells
became quiescent. Similarly, in mouse lymphoblasts (L5178Y) the level of
cAMP did not change significantly when cultures reached stationary phase (328).

Density-dependent inhibited cells can also be induced to initiate DNA replication by infection with polyoma virus or SV40 (see 361), which led Hancock and
Weil (208) to suggest that the "virus activates (or derepresses) a regulatory
system of the host cell which controls initiation of DNA replication." The
release of G$_1$ arrest by viruses is not directly mediated by a component of the
infecting virus. Cells cannot be released from G$_1$ arrest by uv light-irradiated

SV40 or by viral coat proteins (63). Also, the release by SV40 can be prevented by blocking viral metabolism with interferon. Thus, the release of G_1 arrest by viruses is not directly mediated by a component of the infecting virus, but is brought about by expression of the viral genome.

Infection of density-dependent inhibited cells with SV40 leads to a twofold decrease in intracellular cAMP preceding host cell DNA synthesis (405), suggesting that a virus impinges on the regulatory system for host cell DNA replication by affecting cellular cAMP metabolism. In BHK21 cells arrested in G_1, infection with adenovirus type 12 leads to a reduction in cAMP within 8 hours after infection (DNA synthesis begins to rise at 10 hours) and only after viral mRNA is detected (404).

Apparently, replication of the viral DNA of oncogenic DNA viruses is always accompanied by host cell DNA synthesis, suggesting that virus and cell may be subject to the same regulatory influences, i.e., the virus must overcome a regulatory system of the cell before viral DNA synthesis is possible. This idea fits with the discovery that infection of cells in G_1 with SV40 leads to production of viral progeny DNA molecules in the subsequent S period (361). If cells are infected during the S period, no viral DNA is replicated until the S period of the next cell cycle. Thus, in order for SV40 DNA to replicate, it must be present in the cell when replication of the host cell DNA is initiated, that is, both viral and cell DNA are perhaps subject to the same control mechanism for initiation of replication, and this mechanism is turned on only once per cell cycle at the G_1–S border.

If a high intracellular level of cAMP is responsible for the arrest of cells in G_1, presumably the transient reduction in cAMP induced by serum is sufficient to allow cells to pass the point of cAMP sensitivity in G_1 and progress through the remainder of the cell cycle even though the cAMP level returns to a high level. This interpretation is consistent with the observation (315, 540) that exogenous dibutyryl-cAMP does not affect rate of progress of cells through the S period. Addition of dibutyryl-cAMP to cells already in S can lengthen the subsequent G_2 period (315) or even arrest cells in G_2 (348, 540), indicating the presence of a cAMP-sensitive point in G_2.

In general, transformed cells contain less cAMP than their normal counterparts [see, for example (359)]. This is probably the direct result of the observed lowering in adenyl cyclase activity that accompanies transformation (15, 85, 185, 528). Infection of chick embryo fibroblasts with Rous sarcoma virus results in a rapid fall in adenyl cyclase activity, a rapid drop in cAMP (16), and morphological transformation. Infection with a temperature-sensitive mutant of the virus produces the same effects when the cells are kept at 37°C. At 40.5°C, however, the cells are morphologically normal and the level of adenyl cyclase activity is the same as in normal cells. Within 10 minutes of shifting cells from the high temperature to the lower one, the activity of adenyl cyclase drops to one-half, and the level of cAMP soon falls. The morphological change that occurs with shift to the permissive temperature for the mutant virus is prevented by treatment with dibutyryl-cAMP and theophylline (358). Finally, the greater agglutinability of transformed cells with plant lectins compared to normal cells also correlates closely with their lower cAMP content (538, 539).

cAMP may also mediate the inhibitory effect of epinephrine on cell proliferation in epidermal tissue (79). Epinephrine markedly stimulates adenyl cyclase activity of the plasma membrane of erythrocytes (360). Such stimulation is presumably the mechanism by which epinephrine brings about an increased intracellular level of cAMP in hamster epidermis (72). The increased cAMP level induced by epinephrine is accompanied by a decrease in cell proliferation in the epidermal tissue (72). Whether the inhibition of cell proliferation caused by tissue chalones through G_1 or G_2 blocks involves a rise in intracellular cAMP apparently has not been tested.

Finally, the large variableness in the length of G_1 in a homogeneous population of cells (already discussed earlier) may be caused by a transient arrest mediated by cAMP at a checkpoint early in the G_1 period.

REGULATION OF cAMP LEVELS IN CELLS

These various observations and arguments immediately lead in turn to the problem of how intracellular levels of cAMP are regulated. Little is known about this beyond the observations that adenyl cyclase is reduced in virally transformed cells (16, 85). In chick cells transformed by a temperature-sensitive mutant of Rous sarcoma virus, the level of adenyl cyclase drops within 10 minutes after a shift from the nonpermissive (42°C) to the permissive (37°C) temperature (16).

The observation that adenyl cyclase in animal cells is localized in the plasma membrane naturally leads to speculation that the cAMP system is the means by which events at the cell surface (density-dependent inhibition of growth; hormone effects) may modulate cell reproduction (see 82). For example, the surface change that occurs in normal cells in mitosis, in normal cells treated with trypsin, and in normal cells that are transformed is accompanied in each case by a sharply reduced intracellular level of cAMP. Thus, the surface change (to agglutinability) may produce lowered adenyl cyclase activity, reducing cAMP production and allowing cells to escape a restriction mediated by cAMP on traverse through G_1. This does not explain why normal cells in mitosis should undergo the surface change and reduction in cAMP level. Contrary to the hypothesis sketched above, the surface change to agglutinability could conceivably be the effect of lowered intracellular cAMP and not the cause.

cAMP has been implicated in many facets of cell activity, and it seems likely that the molecule must be involved in one or more stages of the coordinated advance (or arrest) of the cell cycle. Some protein kinases activated by cAMP are bound to regulatory molecules when in the inactive state. Activation of the kinase apparently occurs when cAMP binds to the regulatory molecule and dissociates it from the kinase (e.g., 276, 545). Thus, the activation of kinases by cAMP might be modulated during the course of the cell cycle by the cell cycle-dependent synthesis of regulator molecules of the kinases. Kuehn (273) has described a protein kinase in *Physarum* that is *inhibited* by cAMP in a cell cycle-dependent fashion. The inhibitory response of the kinase to cAMP was decreased from mitosis to mid S. For 1 hour during mid S, the kinase was independent of cAMP, and thereafter the capacity of cAMP to inhibit the protein kinase was restored, the completion of restoration coinciding with the termination of DNA synthesis. In these experiments the control of protein kinase was achieved through the coordinate action of cAMP and at least one unidentified factor, conceivably a protein. These observations deal with inhibition of kinase by cAMP rather than activation, but they still provide a model by which cAMP might interact with regulatory proteins to modulate the phosphorylation of other proteins (by protein kinase) involved in cell cycle traverse and the regulation of cell reproduction.

Phosphorylation of nuclear proteins is an early event in the stimulation of proliferation in quiescent lymphocytes by phytohemagglutinin and other mitogens (see 245). Mitogens cause an immediate tenfold rise in cGMP in human lymphocytes (202, 439). cGMP stimulates phosphorylation of nuclear acidic proteins, while an increase in intracellular cAMP inhibits phosphorylation of nuclear acidic proteins (245). Thus, the action of cAMP in G_1 arrest of the cell cycle and of cGMP in the release of this arrest may conceivably be mediated by regulation of phosphorylation of nuclear proteins.

Finally, the possibility that the G_1 arrest of normal tissue cells imposed by chalones may be mediated through cAMP could be examined by measuring the effects of chalones on cultured normal cells and their transformed (tumor) counterparts. Breakdown in control of reproduction of tumor cells may be due in some cases to loss of sensitivity to a chalone. For example, two to four times more lymphocyte chalone is required to inhibit leukemic lymphocyte reproduction compared to normal lymphocytes (236).

11

Nuclear Proteins and the Cell Cycle

Proteins of the nucleus must certainly perform many activities related to the cell cycle, but aside from a few enzymes, little is known about the functions of the bulk of nuclear proteins. Various changes in nuclear proteins during the cell cycle have been measured but the significance of such changes remains obscure because of ignorance about the functions of these proteins.

HISTONES

The most firmly established fact about any of the nuclear proteins is the doubling in histone content concomitant with the synthesis of DNA (Fig. 58). This is a clear case of the synthesis of a specific class of proteins during a specific interval in the cycle, reflecting the activation and inactivation of a set of genes in a temporally controlled fashion [see, for example (88)].

Histone synthesis takes place on small polysomes in the cytoplasm from which completed histone molecules rapidly become bound in the nucleus (371). The absence of detectable mRNA for histones in the cytoplasm of G_1 cells indicates that transcription of histone genes is activated near the G_1–S border. In HeLa blocked by high thymidine, mRNA's for histone synthesis are produced immediately upon release of the block (70). Synthesis of histone mRNA apparently continues through S and G_2. During mitosis or as the cells begin G_1, all histone mRNA is destroyed (8).

Although histone synthesis occurs predominantly during DNA synthesis, degradation and resynthesis of histones can occur in the absence of DNA synthesis in amoeba (393). A low rate of histone synthesis appears to occur in

119

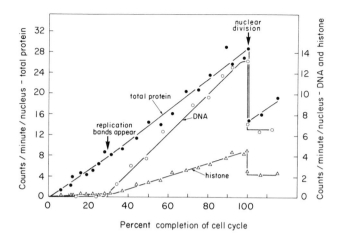

Fig. 58. An example of the synthesis of histones during the S period. Incorporation of tritiated amino acids into total protein (closed circles) and histones (open triangles) was measured for the macronucleus of the ciliated protozoan *Euplotes* during the cell cycle. The S period is defined by the curve for incorporation of ^{3}H-thymidine (open circles). Redrawn from Prescott (391).

G_1-arrested mammalian cells (21, 198, 478) and in cells traversing G_1 (3.5–5% of the rate in S period cells) (198). Uncoupling of DNA and histone syntheses have also been observed in meiotic cells of insects (17, 47) and early amphibian development (7).

Histone Modification

In addition to the timing of histone synthesis considerable information is known about histone phosphorylation (196, 197, 280, 281, 296, 446), acetylation (446), and methylation (50). F_1 histone is not actively phosphorylated in G_1, but phosphorylation takes place in S and G_2. The "superphosphorylation" of F_1 in G_2 is temporally related to chromosome condensation (195). F_1 histone remains phosphorylated during mitosis and is dephosphorylated in G_1 (281, 302). This fits with the finding that F_1 histone is not phosphorylated in nonproliferating cells but is extensively phosphorylated in proliferating cells (28, 29, 353, 365). These several observations suggest that histone F_1 may have a particular role in chromatin structure in mitotic chromosomes (280, 281, 296, 302). Similarly, in *Physarum* the phosphate content of very lysine-rich histone (F_1-like histone) goes up sharply in late G_2, reaching a peak when chromosome condensation is occurring (60–62).

A role of phosphorylation of F_1 histone in chromosome condensation has been strikingly demonstrated by addition of F_1 histone phosphokinase prepared from Ehrlich ascites cells to *Physarum* (60, 61). This phosphokinase specifically catalyzes the phosphorylation of F_1 histone, and when added to the medium, it accelerates the entry of G_2 cells into mitosis (Fig. 59). Obviously, the enzyme is able to enter the plasmodium, perhaps by pinocytosis, and retain activity.

Growing Chinese hamster cells contain two phosphokinases, designated KI and KII, specific for F_1 histone and phosphorylating different sites on F_1 (280). KI is cAMP-dependent and KII is cAMP-independent. The KII phosphokinase appears to be responsible for the "superphosphorylation" of F_1 histone during the G_2-mitosis transition (195, 280).

F_1 is apparently the only histone that does not become methylated (50). Other than these modifications of F_1, measurement of histone modification through the cell cycle has offered little concrete insight into matters of either cell cycle traverse or histone function. For example, acetylations of the various histones follow particular patterns during the cell cycle (446), but the role of histone acetylation in cell cycle traverse is obscure.

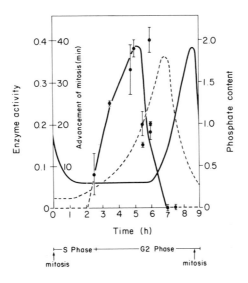

Fig. 59. Acceleration of *Physarum* into mitosis by addition of F_1 histone phosphokinase to the medium. The solid circles indicate the times of addition of enzyme to the medium, and the brackets for each solid circle give the error in the estimate of the advancement of mitosis shown on the left. The dashed line shows normal endogenous enzyme activity, and the solid line shows the phosphate content of F_1 histone. Redrawn from Bradbury *et al.* (61).

NONHISTONE NUCLEAR PROTEINS

The nonhistone nuclear proteins likewise undergo shifts in synthesis and amount in a cell cycle-dependent fashion. Major results of such studies have been reviewed by Stein and Baserga (474).

Synthesis of nonhistone nuclear proteins is continuous through G_1, S, and G_2 (see, for example, Fig. 58). In Chinese hamster cells the rate of synthesis of nonhistone nuclear proteins appears to be high in late G_1 and much lower in S and G_2 (171). Measurements of nuclear proteins in L cells by interference microscopy showed that the amount of such proteins did not increase much during G_1 (552). During the S period nonhistone nuclear proteins accumulated rapidly. Rapid labeling of nonhistone nuclear proteins in G_1 without much increase in amount in the nucleus could be interpreted in two ways: The proteins are synthesized rapidly and degraded at the same rate; or the proteins are synthesized in the cytoplasm where they accumulate and undergo exchange with proteins in the nucleus. Most if not all nuclear proteins are synthesized in the cytoplasm (177), and many nuclear proteins migrate back and forth between nucleus and cytoplasm (176).

Some nonhistone proteins that can be recovered from isolated chromatin of HeLa cells change in their relative proportions through G_1, S, and G_2 (42). One group of proteins decreased just before the beginning of the S period and returned to the original level in mid S.

When cells arrested in G_1 or G_0 are stimulated to proliferate, the amount of nuclear proteins increases greatly before DNA synthesis begins (469). The reactivation of RNA and DNA synthesis in chick erythrocyte nuclei in HeLa cell cytoplasm is preceded by a dramatic flow of nonhistone proteins into the nucleus from the cytoplasm (19, 408). The uptake of proteins by the chick nucleus is particularly fast in HeLa cells in early G_1, a time when the nuclear proteins released to the cytoplasm during mitosis rapidly return to the nucleus (20). Finally, stimulation of proliferation of salivary gland cells in mice by isoproterenol (475) or release of density-dependent inhibition of WI-38 cells with fresh medium (414) is followed almost immediately by a rapid rise in incorporation of labeled amino acids into nonhistone nuclear proteins.

Thus, it appears that changes in amount and synthesis of nonhistone nuclear proteins are related in one way or another to cell proliferation. Possibly some nuclear protein changes are directly involved in progression of the cell cycle. Unfortunately, however, without specific knowledge about the functions of the many nuclear proteins, changes in nuclear proteins during the cell cycle remain largely uninterpretable.

Phosphorylation of nonhistone chromatin proteins occurs continuously during the cell cycle of HeLa cells, although it is markedly reduced during mitosis (380). There are also quantitative differences in phosphorylation of different proteins in G_1, S, G_2, and D. Also, in quiescent BHK cells stimulated to proliferate by transfer to fresh medium, the phosphorylation of nonhistone chromatin proteins increases and reaches a maximum before DNA synthesis starts (117). These various changes presumably reflect changing nuclear activity, but their specific significance is not known.

Some proteins that bind to purified DNA (DNA-binding proteins) are preferentially synthesized at different parts of the cycle in WI-38 cells (97, 98, 189). Conceivably, the cell cycle-dependent synthesis of these proteins is somehow related to progression of the cycle, but without knowledge of the functions of such proteins, interpretations are severely limited.

12

RNA Synthesis and the Cell Cycle

Requirements for RNA synthesis to traverse the G_1, S, and G_2 periods have been noted in previous chapters. In general the cell cycle may proceed for many hours after inhibition of the synthesis of rRNA, but inhibition of synthesis of total RNA is followed fairly quickly by cessation of progress through G_1, S, or G_2. Presumably, this reflects a continuous requirement for mRNA synthesis to drive the cycle.

As mentioned previously the synthesis of all RNA stops during mitosis except for the synthesis of some 4 S and 5 S RNA. In late telophase the synthesis of all classes of nuclear RNA is resumed, and the rate of synthesis rapidly rises to a level that is more or less maintained throughout the G_1 period. Eventually during interphase the rate of total RNA synthesis doubles (553), but reported studies only partially agree on when the increase occurs and whether the increase is sudden or occurs gradually. Zetterberg and Killander (553) found that the rate of RNA synthesis in mouse fibroblasts increases predominantly in the second half of interphase. In HeLa cells the rate appears to increase slowly in G_1 and then rapidly during the S phase (376). Stubblefield et al. (483) suggested from their studies on Chinese hamster cells that the rate of RNA synthesis doubles abruptly in the first part of S, but the data really only show that the rate increases during about the first half of S. The data of Klevecz and Stubblefield (263) more convincingly support the conclusion of a doubling in the rate in early S. The data of Showacre et al. (450) and of Enger and Tobey (130) on Chinese hamster cells indicate a continuous increase in the rate of RNA synthesis through the G_1 and S periods. The results of Scharff and Robbins (430) on HeLa cells also indicate a continuous increase during G_1 and S.

An experiment by Stambrook and Sisken (473) suggests that the differences in rates of RNA synthesis reported in the several experiments described above are probably due to differences in culture conditions. When Chinese hamster V79 cells were cultured in Eagle's minimal essential medium enriched with several kinds of supplements, the rate of ^{3}H-uridine incorporation increased only slightly during G_1 and accelerated sharply as DNA replication began. In a less rich medium the rate of ^{3}H-uridine incorporation in the same cell type increased linearly throughout G_1 and S, with no detectable change in rate at the G_1–S border.

Stimulation of proliferation by medium renewal of cells arrested by density-dependent inhibition is followed within a few minutes by increased RNA synthesis, resulting in considerable increase in total RNA per cell well before DNA synthesis starts (Fig. 60) (189). The increased content of RNA is accompanied by an increased rate of RNA synthesis (Fig. 61) (189) and increased rate of uridine uptake from the medium (112).

A number of attempts have been made at the more difficult analysis of the synthesis of different classes of RNA over the cycle. Clason and Burdon (102) measured synthesis of a large number of low molecular weight fractions of RNA. Their data indicate that these fractions are synthesized most rapidly in mid S or later. However, the synthesis of a variety of stable low molecular weight RNA's in the HeLa cell nucleus appears to bear no particular relationship to the S period (524). According to Pfeiffer's studies (375) the rates of synthesis of nuclear pellet RNA, nuclear supernatant RNA, "mRNA," tRNA, and rRNA all

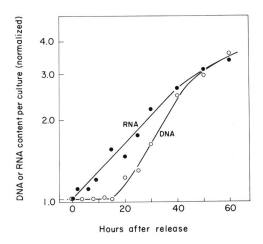

Fig. 60. Increase in RNA in relation to the initiation of DNA synthesis in mouse 3T6 cells following release of arrest by addition of serum. From Green (189).

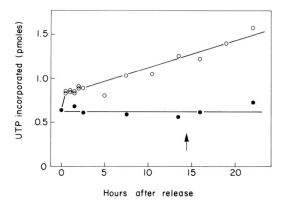

Hours after release

Fig. 61. Rates of RNA synthesis measured by incorporation of labeled uridine triphosphate by ghost 3T6 cells (cell monolayers treated with a neutral detergent, NP-40). Rates (ordinate) were determined from 10-minute labeling periods at intervals after serum stimulation. Open circle, serum-stimulated; closed circle, unstimulated controls. The arrow indicates the beginning of DNA synthesis in stimulated cultures. From Green (189).

increase in the first part of the S phase in HeLa cells. HnRNA is synthesized in G_1, S, and G_2 (362), but the rate of synthesis was not measured. Finally, the synthesis of all classes of RNA occurs throughout the entire interphase in *Physarum* (67). A larger fraction of the genome appears to be transcribed in the S period than in G_2, suggesting differential gene activity during the cycle (147).

Although there are important differences among the foregoing data, we may draw some rather general conclusions. The rate of synthesis of total RNA increases during interphase. The increase cannot be restricted to one point in the cycle since some of the data point to a major increase in G_1, and some point to a major increase in the first part of S, while still other studies indicate that the rate of synthesis increases more or less continuously through interphase. All the major classes of RNA are synthesized continuously throughout interphase.

Regarding the significance of these data on RNA synthesis, they have not yet really provided any important insight into the progression of the cell cycle or its regulation. Undoubtedly, RNA synthesis is crucial to cycle progress, but the proof of this assumption will require better systems of cell synchrony and much finer tools for the analyses of RNA than are now available.

13

Patterns of Enzyme Activities through the Cell Cycle

Changes in the activities of a variety of enzymes have been measured in relation to the cell cycle. In many cases the changes have been shown to be due to degradation and synthesis of the particular enzyme (as opposed to inactivation or activation of existing enzyme molecules). In some cases, e.g., enzymes concerned with production of deoxynucleoside triphosphates, the relevance of the enzymes to progression through the cell cycle is clear, but often it is not, e.g., lactic dehydrogenase, fumarase, and glucose-6-phosphate dehydrogenase.

Mitchison (318) has written an excellent comprehensive review of the information on enzyme patterns available until 1971. In general, in prokaryotes the rate of synthesis of a constitutive enzyme or the potential to synthesize an inducible enzyme doubles at the same time that the relevant genes replicate in the cycle. There are variations in rates of synthesis for some enzymes in prokaryotes that appear to contradict this hypothesis of gene dosage, but these variations are probably due to the imposition of catabolite repression on the pattern of enzyme synthesis during the cycle.

In eukaryotes the patterns of activities for many enzymes do not conform to the relatively simple hypothesis that gene dosage governs the rate (or potential rate in the case of inducible enzymes) of enzyme synthesis. In the yeast *Schizosaccharomyces pombe,* for example, the activities of several enzymes increase in well-defined steps, one step occurring for each enzyme once during each cell cycle. The steps for the different enzymes, however, are spread through the cycle and do not necessarily occur at the time of DNA synthesis (317, 319). The activities of other enzymes in this yeast appear to increase continuously rather than in steps (59, 319). The analyses by Mitchison and Creanor (320) show that for continuously increasing enzymes the rates of increase double at

127

"critical points" in the cycle, but these critical points occur outside of the S period.

Examples of patterns of activities for other enzymes are as follows. In *Euglena* (520) and *Chlorella* (438) the activity per cell of a DNase apparently specific for single-stranded DNA remains low during G_1 and then increases in parallel with the increase in DNA in the cell. The activity of a DNase that attacks native DNA has been shown in *Euglena* (520) and L cells (299) to remain constant through interphase. Presumably, the enzyme activity doubles at the time of cell division. It is not known whether the single-stranded DNase is involved in DNA synthesis, as might be inferred from the results.

In the slime mold *Physarum,* the activity of glucose-6-phosphate dehydrogenase increases continuously through interphase (425), but ribonuclease activity rapidly doubles near the middle of interphase, after DNA synthesis is over (65). The increase in ribonuclease activity apparently represents *de novo* enzyme synthesis since the increase is blocked by cycloheximide. In heteroploid Chinese hamster cells, glucose-6-phosphate dehydrogenase and lactic dehydrogenase activities appear to oscillate, reaching peaks in G_1, early S, and late S (261, 262). In KB cells, Bello (37) found no oscillation in either lactic dehydrogenase or fumarase activities. Various patterns have been demonstrated for acid and alkaline phosphatases, alkaline DNase, succinic dehydrogenase, lactic dehydrogenase, and catalase in synchronized HeLa cells (100, 495). In rat hepatoma cells synchronized by mitotic selection the activities of three enzymes involved in pyrimidine nucleotide synthesis, carbamoylphosphate synthetase, aspartate transcarbamoylase, and orotate phosphoribosyl transferase, increase markedly during G_1, reach peaks during S, and decline sharply during G_2 and mitosis (316).

The synthesis of tyrosine aminotransferase in rat hepatoma cells has been well studied by Tomkins and his colleagues (see 303). The synthesis of the enzyme can be induced by glucocorticoid hormones during the latter part of G_1 and the S period, but not during G_2 or the first few hours of G_1. To explain this and a variety of other observations on the synthesis of tyrosine aminotransferase, a model involving two genes has been proposed which describes the regulation of enzyme synthesis through the cycle: an *S* gene determines the structure of tyrosine aminotransferase, and an *R* gene codes for a labile posttranscriptional repressor of synthesis of the enzyme. In the model both genes are continuously transcribed during the inducible period, i.e., from late G_1 to the beginning of G_2, but in the absence of a glucocorticoid hormone the product of the *R* gene both inhibits the translation and promotes the degradation of the mRNA of the *S* gene. When the inducer is added, it inhibits both actions of the repressor, and mRNA for the enzyme accumulates and is translated. The failure of the enzyme to be induced by glucocorticoids during G_2 and most of G_1 is believed to stem from inhibition of the transcription of both the *S* and the *R* genes. The model is complex, but it appears to be the simplest way to explain the variety of experimental findings.

A few enzyme activities show peak patterns that reflect a burst of rapid synthesis or the activation of the enzyme followed by degradation or inactivation [see Mitchison (318)]. The best-studied examples in eukaryotic cells are thymidine kinase and other enzymes concerned with deoxynucleotide metabolism (see Chapter 4.) These are synthesized *de novo* near the beginning of S. Peak behavior of this sort probably reflects transient derepression of the gene coding for the enzyme, coupled with inherent lability of the enzyme or else its active specific degradation.

So far none of the many demonstrated changes in enzyme activities through the cell cycle provides immediate insight into the problem of the cause and effect continuum that underlies the cell cycle. These studies do, however, support the hypothesis that the progression of the cell cycle is based on a temporally ordered set of transcriptions. The transcriptions are presumably held in sequence by an inductive effect of the translation of one gene product on the transcription of the next. Such an arrangement is perhaps exemplified in the puffing pattern induced by ecdysone in the polytene chromosomes of the dipteran larva (103). Apparently the RNA produced in one RNA puff must be translated in order for induction of RNA synthesis to occur in a second puff.

14

The Genetics of the Cell Cycle

Progression through the cycle depends at least in part on a succession of gene functions. These genes, some of which have been identified by mutation studies, form a set of progression genes that function in the cause and effect progression of the cell cycle. Essentially nothing is yet known about the control mechanisms that maintain the appropriate temporal order of activation of progression genes. Presumably, the product of one gene functions in a way that ultimately leads to activation of the next gene in the sequence. In addition to a set of progression genes, there is strong evidence of the existence of regulatory genes whose functions are to bring about the orderly arrest of the cell cycle in response to nonspecific and specific signals in the environment.

The genetic analysis of the cell cycle is a relatively new area of research that is rapidly becoming a major part of the study of cell reproduction.

HOW MUCH OF THE GENOME IS SPECIFICALLY CONCERNED WITH THE CELL CYCLE?

The genetic analysis of the cell cycle raises the question of how much of the genome is concerned with the events of the cycle in eukaryotic cells. Even approximate estimates are still not possible at the present, but a comparison of the DNA amounts in several eukaryotes is instructive. Laird (278) has pointed out that the sea squirt (*Ciona*) (23) and *Drosophila* (277, 403) contain a haploid amount of DNA that amounts to only about 6% of the amount of DNA present in a haploid mammalian nucleus. There is no reason to suppose that the mammalian cell cycle requires any more genetic information than the cell cycles for cells of *Ciona* and *Drosophila*. It follows therefore that much less than 6% of the DNA in mammalian cells is concerned with cell growth and division. Further, the ciliated protozoan *Oxytricha*, a large (120 × 80 μm) and complex cell, supports all its cellular metabolism, including growth and division on less than 5% of the number of nucleotide sequences in DNA that are present in a mammalian cell (52). It can be tentatively concluded that only a very small fraction of the DNA in a mammalian cell is concerned with the maintenance of basic cellular metabolism as well as all the events that compose the cell life cycle.

GENETIC ANALYSIS OF THE CELL CYCLE IN YEAST

Hartwell and his colleagues (111, 214, 218, 219) have begun the genetic analysis of the cell cycle in yeast *Saccharomyces cerevisiae,* and the results have already provided some fresh insight on functional aspects of the cycle. Yeast cells have several important advantages for the study of cell cycle genetics. They proliferate rapidly on synthetic medium; they are haploid and undergo mating and hence are ideal for mutational work; they have cytological features (Fig. 62) that provide accurate markers for determining the position of a cell in the cycle.

So far, 150 temperature-sensitive (*ts*) mutations affecting 32 genes that control various steps in the cell cycle have been identified. On the basis of these studies Hartwell and his associates (217) have constructed a diagram of events for the yeast cell cycle as shown in Fig. 63. Figure 62 describes the stages of the cell cycle for the budding yeast. Figure 63 indicates the sequential dependence of one event upon another. Initiation of DNA synthesis and bud emergence, for example, are both dependent upon the completion of a prior event in G_1 designated as "Start." By the genetic analysis of the dependence of one event on another Hartwell *et al.* (217) propose that the cell cycle consists of two trains of events that progress separately and which begin at a common Start point in G_1 and then diverge. The two trains of events again connect prior to cytokinesis, i.e., late nuclear division and nuclear migration (see Fig. 63) are both needed for cytokinesis. A new cycle apparently can start without completion of cytokinesis or cell separation. In general, the same is true for eukaryotic cells to the extent that the nuclear cycle of G_1, S, G_2, and nuclear division can proceed without cytokinesis for a number of cycles, resulting in multinucleated cells. In mitotic cells even nuclear division is omitted under some conditions, resulting in endoreduplication (which leads to polyploidy) or polyteny. In yeast the cell cycle genes do not appear to be tightly clustered on the genetic map (216).

Of particular interest in the yeast cycle is the gentically defined point of Start positioned in G_1 some time before the initiation of DNA synthesis. The Start position is probably the normal arrest point of the cycle for yeast cells in stationary phase, for spores, and for cells arrested by the diffusible mating

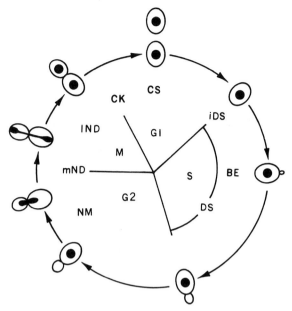

Fig. 62. The life cycle of the yeast cell. CS, cell separation; iDS, initiation of DNA synthesis; BE, bud emergence; DS, DNA synthesis; NM, nuclear migration; mND, medial nuclear division; IND, late nuclear division; CK, cytokinesis (217). Redrawn from L.H. Hartwell, J. Culotti, J.R. Pringle, and B.J. Reid (1974). *Science* **183**, 46–51. Copyright 1974 by the American Association for the Advancement of Science.

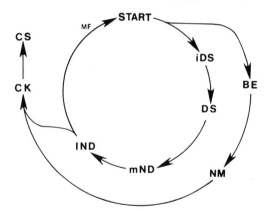

Fig. 63. The dissection of the yeast cell cycle with temperature-sensitive mutants. Events connected by an arrow are proposed to be related such that the occurrence of any given event is dependent upon the completion of the prior event. MF refers to the stage that is sensitive to arrest by mating factor α from cells of an opposite mating type. See Fig. 62 for explanation of abbreviations (217). Redrawn from L.H. Hartwell, J. Culotti, J.R. Pringle, and B.J. Reid (1974). *Science* **183**, 46–51. Copyright (1974) by the American Association for the Advancement of Science.

factor. A temperature-sensitive mutant has been obtained in which the cells are arrested at Start at the restrictive temperature. The mutated gene apparently contributes some function that is essential for the cell to pass the Start and progress through G_1 to the initiation of DNA synthesis. Presumably, the same gene function is repressed by environmental conditions (nutrient depletion or presence of mating factor α) that interrupt the cycle at Start. Start in yeast corresponds to the checkpoint or restriction point in early to mid G_1 postulated to be generally present in the cell cycle for plant and animal cells. Howell and Naliboff (239) have begun a genetic analysis of the cell cycle of *Chlamydomonas reinhardtii* using *ts* mutants. They describe nine such mutants that have block points spread throughout the cell cycle.

GENETIC ANALYSIS OF THE CELL CYCLE
IN MAMMALIAN CELLS

The application of the kind of genetic analysis of the cell cycle that has been started on the yeast cell is in a stage of rapid expansion (496).

Several mutant cell lines have been described that are blocked in G_1 at the non-permissive temperature (usually $39°-41°C$) but traverse G_1 in normal fashion and enter S at a lower temperature. A Chinese hamster cell mutant (K-12), obtained by Roscoe *et al.* (413), is able to initiate DNA synthesis at $37°C$ but not at $40°C$ (Fig. 64) (461, 462). Cells shifted from $37°$ to $40°C$ at 0, 2, or 5 hours after mitosis are blocked in DNA synthesis while cells shifted from $40°$ to $37°C$ at 0, 2, or 5 hours after mitosis enter the S period on schedule. Cells shifted from $37°$ to

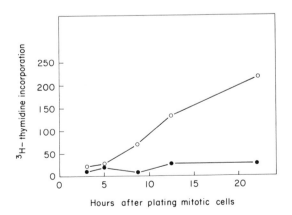

Fig. 64. A temperature-sensitive mutant of Chinese hamster K-12 cells unable to initiate DNA synthesis. Mitotically selected cells were plated at $37°C$ (open circles) or $40°C$ (closed circles) and incubated with ^{3}H-thymidine. From Smith and Wigglesworth (461).

$40°C$ while they are arrested with hydroxyurea (probably arrested in very early S), synthesize DNA (at $40°C$) when the hydroxyurea is removed. Hence, the temperature-sensitive block appears to be in late G_1 between 5 hours after mitosis and the hydroxyurea blockpoint. Since the cells lose viability after 12 hours at the nonpermissive temperature, it is likely that the mutation does not block cells at the normal arrest point in G_1. Roscoe et al. (413) have analyzed this hamster mutant cell line and concluded that the arrest point may be located several hours before initiation of S. When an asynchronous culture is shifted from $33°$ to $40°C$, DNA synthesis occurs at a normal rate for 3 to 4 hours, suggesting that cells in the last 3 to 4 hours of G_1 enter DNA synthesis at the nonpermissive temperature. The rate of mitosis continues normally for 15 hours at $40°C$. Since $S + G_2 = 11$ to 12 hours at $40°C$, the falloff of mitosis is presumed to be due to block of cells 3 to 4 hours back in G_1. At the nonpermissive temperature this mutant undergoes a decrease in some of its nuclear phosphoprotein (407a), but there is no way to tell whether this change is the cause of G_1 arrest or the result of G_1 arrest.

In a mutant selected from a near-diploid mouse cell line by Liskay (290) G_1 cells fail to enter the S period when shifted from $33°$ to $39°C$. Cells already in S at the time of shift to the nonpermissive temperature traverse the cell cycle, complete mitosis, and move into the next G_1 period. The defect is presumably in late G_1.

The ts mutant derived from the BHK cell line by Burstin et al. (87) is also blocked in G_1 at the nonpermissive temperature. At the nonpermissive temperature the amount of cAMP per cell goes up, and the cells gradually acquire the G_1 amount of DNA. When cells blocked with hydroxyurea at the permissive temperature are raised to the nonpermissive temperature in the absence of hydroxyurea, they synthesize DNA. Hence the mutant block point precedes the hydroxyurea blockpoint. Cells arrested in G_1 by serum deprivation and shifted to the nonpermissive temperature at the time of release by serum addition do not enter S. Cells blocked by isoleucine deprivation and shifted to the nonpermissive temperature at the same time as they are released by isoleucine addition subsequently do enter DNA synthesis. Hence the mutant block point occurs earlier in G_1 than the isoleucine arrest point and occurs at or after the point in G_1 at which serum deprivation arrests cells.

Smith and Wigglesworth (463) have selected from a Syrian hamster cell line a ts mutant for cytokinesis. At the nonpermissive temperature $(39°C)$ cells undergo mitosis normally, but many fail to divide, giving rise to binucleated cells. A similar mutant of Chinese hamster fibroblasts fails to undergo cytokinesis at $39°C$, and multinucleated cells are produced (222). Cytokinesis occurs at the permissive temperature $(34°C)$ although some residual expression of the mutation is evident from the presence of some binucleated cells. Wang (521) has obtained a *ts* mutant of hamster cells that is blocked in metaphase at the nonpermissive temperature.

THE CONCEPT OF REGULATORY GENES
FOR CELL REPRODUCTION

Although essentially nothing is known about those events in G_1 which are involved in the arrest of the cycle and the control of cell reproduction, some insight into the matter is provided by those situations in which the usual G_1 arrest has become less efficient. The consequent loss of regulation of cell reproduction is clearly recognizable in multicellular organisms as neoplastic disease. In cultures of normal animal cells the loss of regulation of cell reproduction may be measured as the loss of density-dependent inhibition of cell reproduction. It is usually assumed that density-dependent inhibition of cell reproduction in cultured normal cells represents some remnant of the mechanism by which cell reproduction is regulated within an organism. It is then further assumed that the change within a cell that underlies the loss of regulation of cell reproduction in culture is the same change that underlies the development of neoplasia within an organism. The assumption that the loss of regulation has the same molecular basis both in cultured cells and in the organisms is supported by several kinds of evidence, and it is accepted practice to describe the cellular change in both situations as transformation. For example, the obliteration of density-dependent inhibition of growth by transformation of cells in culture with various oncogenic viruses also is manifested as a loss of regulation of reproduction in those same cells when they are reimplanted into an organism. Moreover, the property of density-dependent inhibition of growth may be lost by degrees, and the degree of loss is accompanied by a roughly corresponding increase in the malignancy of those cells when implanted into a whole animal (1, 382, 383). There are other parallels between loss of regulation in culture and in the organism, but for present purpose perhaps the most important is that both situations are characterized by an impairment in the mechanism by which the cell is retained in G_1.

Much evidence indicates that the impairment of the arrest mechanism in G_1 is the result of a genetic change within the cell. Although not yet conclusively proved, a variety of experimental findings supports the idea that radiation causes cellular transformation by the induction of mutations or induction of viral transformation. Many chemical carcinogens are proved mutagens, and possibly all chemical carcinogens are mutagens. Ames *et al.* (12), for example, have shown that 18 different aromatic-type chemical carcinogens can be activated by liver homogenate into forms that act as frameshift mutagens affecting histidine synthesis in *Salmonella.* Furthermore, a large body of experimental work indicates that *integration* of the viral genome or part of the genome into a cellular chromosome is required for transformation of a cell. As yet, it is not known how this genetic change in a cell, in the form of the addition of part or all of the viral genome, results in the alteration of the G_1 arrest mechanism of the cell. This capacity of oncogenic viruses to override regulation of the cell cycle appears to reside in a single viral gene (183, 304, 357). Apparently, integration of this viral gene modifies the transcriptional program and hence the phenotype of the cell in a way that, among other things, lessens the ability of cells to be held in G_1. Finally, in some neoplastic growths it is clear that the loss of G_1 arrest stems from mutation(s) of the cellular genome (see below).

The evidence that contact between cells is a necessary part of density-dependent inhibition of growth suggests that the genetic change underlying transformation results in an essential change in the cell surface. The occurrence of an essential membrane change has been demonstrated in a number of kinds of experiments with viral transformation, particularly by the observations of Burger and his colleagues on agglutination with plant lectins (see 149). Studies with temperature-sensitive mutants of polyoma virus suggest that the change in the cell surface in the course of transformation is coded for by a viral gene (39, 124). Whether this virus activity is the one that overrides regulation of cell reproduction in G_1 or whether other viral genes are involved is not yet clear. The activation of cellular DNA synthesis by SV40 requires expression of the viral genome since the viral effect can be blocked by interferon and since uv-inactivated virus or viral coat protein are ineffective (63).

Another particularly straightforward example of a genetic basis for the failure of regulation of cell reproduction has been described in *Drosophila* by Gateff and Schneiderman (163). In the presence of a homozygous recessive mutation at locus *l(2)gl(4)* (on the second chromosome) the brain anlage in the larva fails to differentiate normally, and the neuroblasts continue to reproduce beyond the normal limit. These neuroblasts also continue to reproduce when transplanted to an adult fly, eventually killing the host through overgrowth. The neuroblastoma is invasive and lethal and can be serially transplanted from one adult to another. The gene locus in question must normally play an essential role in the development of the regulation of cell reproduction during the differentiation of neural tissue within the larva. Thus, the gene in question may be properly called a regulatory gene that is needed for the control of the reproduction of neural cells in the larva.

The idea of regulatory genes for cell reproduction in animals was, in fact, established much earlier in the little-recognized works of Gordon (179) and Kosswig (267) on hybrids between two tropical fish known as the platy and the swordtail. The wild-type platy contains macromelanophores in its dorsal fin and has genes in its genetic makeup for controlling the reproduction of the melanophores. The swordtail lacks the dorsal fin melanophores and, naturally enough, lacks the genes for regulating macromelanophore reproduction. In F_1 hybrids the reproduction of macromelanophores is increased somewhat, but in the progeny from a backcross between the F_1 hybrid and a swordtail, macromelanophore reproduction is intense and results in lethal invasive melanomas. Careful genetic analysis has shown that the development of melanomas is due to the dilution in hybrids of genes required to regulate macromelanophore reproduction.

There is also substantial evidence from studies on hybrid crosses of plants, particularly for *Nicotiana* species, for the existence of genes that regulate cell reproduction [see review by Smith (466)].

These observations lead to the concept of a large family of regulatory genes, individual members of which are specific for regulation of cell reproduction in different kinds of differentiated cells. Thus, it is proposed that a regulatory gene(s) is among the genes brought into play in the differentiation of any particular specialized cell to control the reproduction of that cell type. The mutation, deletion, or alteration in activity of such regulatory genes is a reasonable model for explaining the loss of reproduction control, whether induced by radiation, a chemical carcinogen, an oncogenic virus, or mutation. On the other hand we have no specific clues about the means by which the actions of regulatory genes interconnect either with properties of the cell surface or with other facets of the cell to produce regulation of reproduction. Some observations suggest that adenyl cyclase in the cell surface could influence the activity of regulatory genes, perhaps through the effects of cyclic AMP on the phosphorylation of proteins by cAMP-dependent protein kinases.

Regulatory genes are presumed to contribute functions that repress (and thereby regulate) cell reproduction. This view of regulatory gene action is at least consistent with demonstrations in cell fusion studies that the presence of regulation of reproduction can be dominant over the absence of regulation, i.e., in fusions between one cell with and one without regulation, the hybrid cell has the property of regulation. This dominance of the regulation property is measured both as suppression of tumor malignancy in hybrid cells (69, 213, 260, 531–533) and as the presence of density-dependent inhibition of growth (529).

Dominance of the regulatory property in these several fusion studies could be explained if transformation has occurred through mutation, deletion, or inactivation of both copies of a regulatory gene in a diploid cell. Fusion to a cell expressing regulation would therefore be expected to produce a hybrid cell in which the property of regulation would be imposed on the transformed genome. In contrast to the foregoing, hybrids formed by fusing a polyoma-transformed mouse cell with a normal mouse cell (115) or an SV40-transformed human cell with a normal human cell (110) behave as transformed cells. Thus, we are confronted with the possibility that the transformed state produced by a virus is dominant, presumably because a viral gene product overrides the cell's regulatory system, while the transformed state produced by mutation, deletion, or inactivation of regulatory genes would be recessive. Such genetic recessiveness for the transformed state is exemplified by the neuroblastoma in *Drosophila* and melanoma in fish hybrids discussed earlier.

It is conceivable that the mechanism by which cell reproduction is regulated may be made up of several steps, each of which is based on a different gene. Thus, regulation may depend on the actions of a set of genes rather than a single gene. This is suggested by an experiment of Levisohn and Thompson (286) in which two malignant cells lacking density-dependent inhibition of growth were fused to produce a cell hybrid that possessed density-dependent inhibition. The apparent complementarity says that the expression of density-dependent inhibition of growth requires the function of more than one gene.

The complexity of genetic regulation of reproduction is also illustrated by temperature-sensitive mutants of rat cells that had been transformed by a chemical carcinogen (544). The *ts* mutant cells grow in agar suspension at $36°C$ but not at $40°C$, while the transformed parent grows at both temperatures. Monolayer culture of the mutant appears to be density-dependent inhibited at $40°C$ but not at $36°C$. In summary, the cells were originally transformed by a chemical carcinogen, presumably by a mutational event. A second mutation causes the cells to revert to "normal" when grown at $40°C$ but not at $36°C$. Presumably, the first mutation, resulting in transformation, occurred in a regulatory gene. The second mutation somehow reinstates the lost regulatory gene function at $40°C$ but not at $36°C$.

Both cold- and heat-sensitive mutations have been obtained starting with spontaneously transformed lung cells of the Chinese hamster (325). After the transformed cells were exposed to a chemical mutagen, cells were selected that still behaved as transformed cells at $39.5°C$ but adopted the growth characteristics of normal cells at $34.5°C$ (cold-sensitive mutants). Cells were also selected that had normal growth characteristics at $39.5°C$ but grew like transformed cells at $34.5°C$ (heat-sensitive mutants). These results and those described above for rat cells are difficult to assimilate into a simple hypothesis of regulation by regulatory genes.

In summary, a variety of kinds of evidence points to the presence of genes that regulate the rate of cell reproduction, and when these regulatory genes are activated by various environmental conditions (presence of a specific chalone, nutrient deprivation, density-dependent inhibition, low serum, etc.), the cell is retained in the G_1 period (shunted in G_0). Such a model must of course also account for the many degrees of regulation within an organism, from slight retardation to irreversible shutdown of cell reproduction. It must in addition include the flexibility that allows cells to change their rate of reproduction drastically in response to appropriate environmental stimuli (hormones, tissue injury, foreign antigens, increased functional load, etc.).

FINAL COMMENT

Some of the main points of this review are summarized in the diagram of the cell life cycle shown in Fig. 65. In addition to the four main sections, the temporal positions of a variety of other events are marked. To avoid making the diagram even more cluttered, some events of the cell cycle discussed in this review have not been included. For an explanation of the more cryptic labels in the diagram the reader should consult various chapters. The events marked in Fig. 65 are components of the cause and effect continuum of gene expressions that must underlie the progression of the cycle. Cell cycle continuity is considered to be based on the sequential transcription and translation of the set of cell

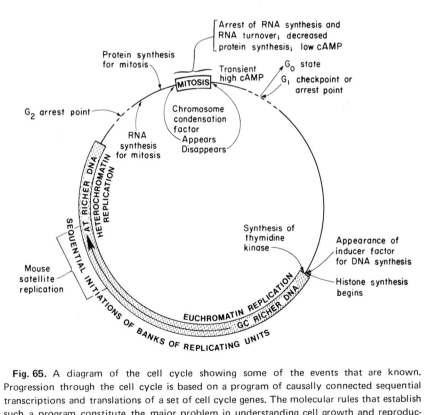

Fig. 65. A diagram of the cell cycle showing some of the events that are known. Progression through the cell cycle is based on a program of causally connected sequential transcriptions and translations of a set of cell cycle genes. The molecular rules that establish such a program constitute the major problem in understanding cell growth and reproduction.

cycle genes that are held in temporal order by the activation of one cell cycle gene by the function of an antecedently activated gene. Cell proliferation is regulated primarily by interruption of the G_1 period several hours before the initiation of the S period, although cells occasionally may be arrested in the G_2 period. Interruption of G_1 traverse is probably achieved by a cell cycle regulatory gene acting in early G_1. Understanding regulation of cell proliferation and loss of regulation will depend upon discovery of the molecular events that underlie the G_1 period, particularly those G_1 events concerned with the G_1 arrest of the cell cycle.

Bibliography

The numbers in brackets following the references indicate the page numbers on which the references are cited.

1. Aaronson, S.A., and Todaro, G.J. (1968). Basis for the acquisition of malignant potential by mouse cells cultivated *in vitro. Science* **162**, 1024–1026. [135]
2. Abell, C.W., and Monahan, T.M. (1973). The role of adenosine 3′,5′-cyclic monophosphate in the regulation of mammalian cell division. *J. cell Biol.* **59**, 549–558. [107, 112]
3. Abraham, K.A., Pryme, I.F., Åbro, A., and Dowben, R.M. (1973). Polysomes in various phases of the cell cycle in synchronized plasmacytoma cells. *Exp. Cell Res.* **82**, 95–102. [97]
4. Abramova, N.B., and Neyfakh, A.A. (1973). Migration of newly synthesized RNA during mitosis. III. Nuclear RNA in the cytoplasm of metaphase cells. *Exp. Cell Res.* **77**, 136–142. [95]
5. Adams, R.L.P. (1969). Phosphorylation of tritiated thymidine by L929 mouse fibroblasts. *Exp. Cell Res.* **56**, 49–54. [55]
6. Adams, R.L.P., Abrams, R., and Lieberman, I. (1966). Deoxycytidylate synthesis and entry into the period of deoxyribonucleic acid replication in rabbit kidney cells. *J. Biol. Chem.* **241**, 903–905. [57]
7. Adamson, E.D., and Woodland, H.R. (1974). Histone synthesis in early amphibian development: Histone and DNA syntheses are not co-ordinated. *J. Mol. Biol.* **88**, 263–285. [120]
8. Allred, L.E., Arlinghaus, R.B., and Humphrey, R.M. (1976). Histone-like messenger ribonucleoprotein particles. In preparation. [119]
9. Alpen, E.L., and Johnston, M.E. (1967). DNA synthetic rate and DNA content of nucleated erythroid cells. *Exp. Cell Res.* **47**, 177–192. [51]
10. Amaldi, F., Buongiorno-Nardelli, M., Carnevali, F., Leoni, L., Mariotti, D., and Pomponi, M. (1973). Replicon origins in Chinese hamster cell DNA. II. Reproducibility. *Exp. Cell Res.* **80**, 79–87. [75]
11. Amaldi, F., Giacomoni, D., and Zito-Bignami, R. (1969). On the duplication of ribosomal RNA cistrons in Chinese hamster cells. *Eur. J. Biochem.* **11**, 419–423. [82]
12. Ames, B.N., Durston, W.E., Yamasaki, E., and Lee, F.D. (1973). Carcinogens are mutagens: A simple test system combining liver homogenates for activation and bacteria for detection. *Proc. Natl. Acad. Sci. U.S.A.* **70**, 2281–2285. [135]

13. Andersen, H.A., and Engberg, J. (1975). Timing of the ribosomal gene replication in *Tetrahymena pyriformis. Exp. Cell Res.* **92**, 159–163. [82]

14. Anderson, E.C., Bell, G.I., Petersen, D.F., and Tobey, R.A. (1969). Cell growth and division. IV. Determination of volume growth rate and division probability. *Biophys. J.* **9**, 246–263. [28]

15. Anderson, W.B., Gallo, M., and Pastan, I. (1974). Adenylate cyclase activity in fibroblasts transformed by Kirsten or Moloney sarcoma viruses. *J. Biol. Chem.* **249**, 7041–7048. [116]

16. Anderson, W.B., Johnson, G.S., and Pastan, I. (1973). Transformation of chick-embryo fibroblasts by wild-type and temperature-sensitive Rous sarcoma virus alters adenylate cyclase activity. *Proc. Natl. Acad. Sci. U.S.A.* **70**, 1055–1059. [114, 116–117]

17. Antropova, E.N., and Bogdanov, Yu.F. (1970). Cytophotometry of DNA and histone in meiosis of *Pyrrhocoris apterus. Exp. Cell Res.* **60**, 40–44. [120]

18. Aoki, Y., and Moore, G.E. (1970). Comparative study of mitotic stages of cells derived from human peripheral blood. *Exp. Cell Res.* **59**, 259–266. [47]

19. Appels, R., Tallroth, E., Appels, D.M., and Ringertz, N.R. (1975). Differential uptake of protein into the chick nuclei of HeLa × chick erythrocyte heterokaryons. *Exp. Cell Res.* **92**, 70–78. [122]

20. Appels, R., Bell, P.B., and Ringertz, N.R. (1975). The first division of HeLa × chick erythrocyte heterokaryons. Transfer of chick nuclei to daughter cells. *Exp. Cell Res.* **92**, 79–86. [122]

21. Appels, R., and Ringertz, N.R. (1974). Metabolism of F1 histone in G_1 and G_0 cells. *Cell Differ.* **3**, 1–8. [120]

22. Arrighi, F.E., and Hsu, T.C. (1965). Experimental alteration of metaphase chromosome morphology. Effect of actinomycin D. *Exp. Cell Res.* **39**, 305–308. [89]

23. Atkin, N.B., and Ohno, S. (1967). The comparative DNA content of 19 species of placental mammals, reptiles, and birds. *Chromosoma* **17**, 1–10. [130]

24. Augenlicht, L.H., and Baserga, R. (1974). Changes in the G0 state of WI-38 fibroblasts at different times after confluence. *Exp. Cell Res.* **89**, 255–262. [45]

25. Aujard, C., Chany, E., and Frayssinet, C. (1973). Inhibition of DNA synthesis of synchronized cells by liver extracts acting in G_1 phase. *Exp. Cell Res.* **78**, 476–478. [44]

26. Avanzi, S., Brunori, A., and D'Amato, F. (1969). Sequential development of meristems in the embryo of *Triticum durum.* A DNA autoradiographic and cytophotometric analysis. *Dev. Biol.* **20**, 368–377. [48]

27. Balazs, I., and Schildkraut, C.L. (1971). DNA replication in synchronized cultured mammalian cells. II. Replication of ribosomal cistrons in thymidine-synchronized HeLa cells. *J. Mol. Biol.* **57**, 153–158. [82]

28. Balhorn, R., Chalkley, R., and Granner, D. (1972). Lysine-rich histone phosphorylation. A positive correlation with cell replication. *Biochemistry* **11**, 1094–1098. [120]

29. Balhorn, R., Rieke, W.O., and Chalkley, R. (1971). Rapid electrophoretic analysis for histone phosphorylation. A reinvestigation of phosphorylation of lysine-rich histone during rat liver regeneration. *Biochemistry* **10**, 3952–3959. [120]

30. Bannai, S., and Sheppard, J.R. (1974). Cyclic AMP, ATP and cell contact. *Nature (London)* **250**, 62–64. [113]

31. Barlow, P.W., and Macdonald, P.D.M. (1973). An analysis of the mitotic cell cycle in the root meristem of *Zea mays. Proc. R. Soc. London, Ser. B* **183**, 385–398. [47]

32. Baserga, R., Rovera, G., and Farber, J. (1971). Control of cellular proliferation in human diploid fibroblasts. *In Vitro* **7**, 80–87. [44]

33. Baserga, R., and Wiebel, F. (1969). The cell cycle of mammalian cells. *Int. Rev. Exp. Pathol* **7**, 1–25. [44]

34. Baserga, R. (1963). Mitotic cycle of ascites tumor cells. *Arch. Pathol.* **75**, 156. [51]

35. Bassleer, R. (1968). Contribution to the study of nuclear total proteins and DNA during the mitotic cycle in fibroblasts cultivated *in vitro* and in Ehrlich ascites cells. *Histochemie* **14**, 89–102. [94]

36. Bello, L.J. (1974). Regulation of thymidine kinase synthesis in human cells. *Exp. Cell Res.* **89**, 263–274. [56]

37. Bello, L.J. (1969). Studies on gene activity in synchronized cultures of mammalian cells. *Biochim. Biophys. Acta* **179**, 204–213. [128]

38. Benbow, R.M., and Ford, C.C. (1975). Cytoplasmic control of nuclear DNA synthesis during early development of *Xenopus laevis:* A cell-free assay. *Proc. Natl. Acad. Sci. U.S.A.* **72**, 2437–2441. [67]

39. Benjamin, T.L., and Burger, M.M. (1970). Absence of a cell membrane alteration function in non-transforming mutants of polyoma virus. *Proc. Natl. Acad. Sci. U.S.A.* **67**, 929–934. [136]

40. Bennett, L.L., Jr., Smithers, D., and Ward, C.T. (1964). Inhibition of DNA synthesis in mammalian cells by actidione. *Biochim. Biophys. Acta* **87**, 60–69. [83]

41. Berezney, R., and Coffey, D.S. (1975). Nuclear protein matrix: Association with newly synthesized DNA. *Science* **189**, 291–293. [72]

42. Bhorjee, J.S., and Pederson, T. (1972). Nonhistone chromosomal proteins in synchronized HeLa cells. *Proc. Natl. Acad. Sci. U.S.A.* **69**, 3345–3349. [122]

43. Bjursell, G., and Reichard, P. (1973). Effects of thymidine on deoxyribonucleoside triphosphate pools and deoxyribonucleic acid synthesis in Chinese hamster ovary cells. *J. Biol. Chem.* **248**, 3904–3909. [32–33]

44. Blenkinsopp, W.K. (1969). Cell proliferation in the epithelium of the oesophagus, trachea and ureter in mice. *J. Cell Sci.* **5**, 393–401. [47]

45. Blondel, B. (1968). Relation between nuclear fine structure and ^{3}H-thymidine incorporation in a synchronous cell culture. *Exp. Cell Res.* **53**, 348–356. [68]

46. Blumenthal, A.B., Kriegstein, H.J., and Hogness, D.S. (1974). The units of DNA replication in *Drosophila melanogaster* chromosomes. *Cold Spring Harbor Symp. Quant. Biol.* **38**, 205–223. [78]

47. Bogdanov, Yu.F., Liapunova, N.A., Sherudilo, A.I., and Antropova, E.N. (1968). Uncoupling of DNA and histone synthesis prior to phophase I of meiosis in the cricket *Grillus (Acheta) domesticus* L. *Exp. Cell Res.* **52**, 59–70. [120]

48. Bombik, B.M., and Burger, M.M. (1973). c-AMP and the cell cycle: Inhibition of growth stimulation. *Exp. Cell Res.* **80**, 88–94. [114]

49. Bootsma, D., Budke, L., and Vos, O. (1964). Studies on synchronous division of tissue culture cells initiated by excess thymidine. *Exp. Cell Res.* **33**, 301–309. [32]

50. Borun, T.W., Pearson, D., and Paik, W.K. (1972). Studies of histone methylation during the HeLa S-3 cell cycle. *J. Biol. Chem.* **247**, 4288–4298. [120–121]

51. Bosmann, H.B. (1971). Mitochondrial biochemical events in a synchronized mammalian cell population. *J. Biol. Chem.* **246**, 3817–3823. [63]

52. Bostock, C.J., and Prescott, D.M. (1972). Evidence of gene diminution during the formation of the macronucleus in the protozoan, *Stylonychia. Proc. Natl. Acad. Sci. U.S.A.* **69**, 139–142. [131]

53. Bostock, C.J., Prescott, D.M., and Hatch, F.T. (1972). Timing of replication of the satellite and main band DNAs in cells of the kangaroo rat (*Dipodomys ordii*). *Exp. Cell Res.* **74**, 487–495. [81]

54. Bostock, C.J., and Prescott, D.M. (1971). Buoyant density of DNA synthesized at different stages of the S phase of mouse L-cells. *Exp. Cell Res.* **64**, 267–274. [80–81]

55. Bostock, C.J., and Prescott, D.M. (1971). Buoyant density of DNA synthesized at different stages of S phase in Chinese hamster cells. *Exp. Cell Res.* **64**, 481–484. [80]

56. Bostock, C.J., and Prescott, D.M. (1971). Shift in buoyant density of DNA during the

synthetic period and its relation to euchromatin in mammalian cells. *J. Mol. Biol.* **60**, 151–162. [80–81]

57. Bostock, C.J., Prescott, D.M., and Kirkpatrick, J.B. (1971). An evaluation of the double thymidine block for synchronizing mammalian cells at the G_1–S border. *Exp. Cell Res.* **68**, 163–168. [32–33]

58. Bostock, C.J. (1970). DNA synthesis in the fission yeast *Schizosaccharomyces pombe*. *Exp. Cell Res.* **60**, 16–26. [9, 49]

59. Bostock, C.J., Donachie, W.D., Masters, M., and Mitchison, J.M. (1966). Synthesis of enzymes and DNA in synchronous cultures of *Schizosaccharomyces pombe*. *Nature (London)* **210**, 808–810. [127]

60. Bradbury, E.M., Inglis, R.J., and Matthews, H.R. (1974). Control of cell division by very lysine rich histone (F1) phosphorylation. *Nature (London)* **247**, 257–261. [120–121]

61. Bradbury, E.M., Inglis, R.J., Matthews, H.R., and Langan, T.A. (1974). Molecular basis of control of mitotic cell division in eukaryotes. *Nature (London)* **249**, 553–556. [120–121]

62. Bradbury, E.M., Inglis, R.J., Matthews, H.R., and Sarner, N. (1973). Phosphorylation condensation. *Eur. J. Biochem.* **33**, 131–139. [120]

63. Brandner, G., Boehlandt, D., Burger, J., and Leveringhaus, M. (1971) Induction of cellular DNA synthesis during lytic infection with SV40: A function of viral genome. *Arch. Gesamte Virusforsch.* **34**, 323–331. [116, 136]

64. Braun, R., and Rüedi-Wili, H. (1971). Early replicating DNA of *Physarum* is denser than late replicating DNA. *Experientia* **27**, 1412. [80]

65. Braun, R., and Behrens, K. (1969). A ribonuclease from *Physarum*. Biochemical properties and synthesis in the mitotic cycle. *Biochim. Biophys. Acta* **195**, 87–98. [55, 128]

66. Braun, R., and Wili, H. (1969). Time sequence of DNA replication in *Physarum*. *Biochim. Biophys. Acta* **174**, 246–252. [79]

67. Braun, R., Mittermayer, C., and Rusch, H.P. (1966). Sedimentation patterns of pulse-labeled RNA in the mitotic cycle of *Physarum polycephalum*. *Biochim. Biophys. Acta* **114**, 27–35. [126]

68. Braun, R., Mittermayer, C., and Rusch, H.P. (1965). Sequential temporal replication of DNA in *Physarum polycephalum*. *Proc. Natl. Acad. Sci. U.S.A.* **53**, 924–931. [79]

69. Bregula, U., Klein, G., and Harris, H. (1971). The analysis of malignancy by cell fusion. II. Hybrids between Ehrlich cells and normal diploid cells. *J. Cell Sci.* **8**, 673–680. [137]

70. Breindl, M., and Gallwitz, D. (1973). Identification of histone messenger RNA from HeLa cells. *Eur. J. Biochem.* **32**, 381–391. [119]

71. Brent, T.P., Butler, J.A.V., and Crathorn, A.R. (1965). Variations in phosphokinase activities during the cell cycle in synchronous populations of HeLa cells. *Nature (London)* **207**, 176–177. [55]

72. Brønstad, G.O., Elgjo, K., and Øye, I. (1971). Adrenaline increases cyclic 3'5'-AMP formation in hamster epidermis. *Nature (London), New Biol.* **233**, 78–79. [117]

73. Brown, J.M. (1968). Long G1 or G0 state: A method of resolving the dilemma for the cell cycle of an *in vivo* population. *Exp. Cell Res.* **52**, 565–570. [47]

74. Brunner, M. (1973). Regulation of DNA synthesis by amino acid limitation. *Cancer Res.* **33**, 29–32. [54]

75. Brunori, A., and D'Amato, F. (1967). The DNA content of nuclei in the embryo of dry seeds of *Pinus pinea* and *Lactuca sativa*. *Caryologia* **20**, 153–161. [48–49]

76. Bryant, T.R. (1969). DNA synthesis and cell division in germinating onion. II. Mitotic cycle and DNA content. *Caryologia* **22**, 139–148. [48]

77. Bryant, T.R. (1969). DNA synthesis and cell division in germinating onion. I. Onset of DNA synthesis and mitosis. *Caryologia* **22**, 127–138. [48]

78. Bücking-Throm, E., Duntze, W., Hartwell, L.H., and Manney, T.R. (1973). Reversible arrest of haploid yeast cells at the initiation of DNA synthesis by a diffusible sex factor. *Exp. Cell Res.* **76**, 99–110. [5]

79. Bullough, W.S., and Laurence, E.B. (1968). The role of glucocorticoid hormones in the control of epidermal mitosis. *Cell Tissue Kinet.* **1**, 5. [117]

80. Bullough, W.S., Hewett, C.L., and Laurence, E.B. (1964). The epidermal chalone: A preliminary attempt at isolation. *Exp. Cell Res.* **36**, 192–200. [48]

81. Bullough, W.S., and Laurence, E.B. (1964). Mitotic control by internal secretion: The role of the chalone–adrenalin complex. *Exp. Cell Res.* **33**, 176–194. [48]

82. Burger, M.M. (1973). Surface changes in transformed cells detected by lectins. *Fed. Proc., Fed. Am. Soc. Exp. Biol.* **32**, 91–101. [104, 106–107, 117]

83. Burger, M.M., Bombik, B.M., Breckenridge, B. McL., and Sheppard, J.R. (1972). Growth control and cyclic alterations of cyclic AMP in the cell cycle. *Nature (London), New Biol.* **239**, 161–163. [107–108, 114]

84. Bürk, R.R. (1970). One-step growth cycle for BHK21/13 hamster fibroblasts. *Exp. Cell Res.* **63**, 309–316. [44, 51]

85. Bürk, R.R. (1968). Reduced adenyl cyclase activity in a polyoma virus transformed cell line. *Nature (London)* **219**, 1272–1275. [116–117]

86. Burns, E.R. (1971). Synchronous and asynchronous DNA synthesis in multinucleated Ehrlich ascites tumor cells compared with multinucleated cells cultured from frog lung. *Exp. Cell Res.* **66**, 152–156. [62]

87. Burstin, S.J., Meiss, H.K., and Basilico, C. (1974). A temperature-sensitive cell cycle mutant of the BHK cell line. *J. Cell. Physiol.* **84**, 397–408. [134]

88. Butler, W.B., and Mueller, G.C. (1973). Control of histone synthesis in HeLa cells. *Biochim. Biophys. Acta* **294**, 481–496. [119]

89. Cairns, J. (1966). Autoradiography of HeLa cell DNA. *J. Mol. Biol.* **15**, 372–373. [74–75]

90. Callan, H.G. (1972). Replication of DNA in the chromosomes of eukaryotes. *Proc. R. Soc. London, Ser. B.* **181**, 19–41. [75–76]

91. Callan, H.G., and Taylor, J.H. (1968). A radioautographic study of the time course of male meiosis in the newt *Triturus vulgaris. J. Cell Sci.* **3**, 615–626. [76]

92. Cameron, I.L. (1966). A periodicity of tritiated-thymidine incorporation into cytoplasmic deoxyribonucleic acid during the cell cycle of *Tetrahymena pyriformis. Nature (London)* **209**, 630–631. [63]

93. Cameron, I.L., and Padilla, G.M., eds. (1966). "Cell Synchrony: Studies in Biosynthetic Regulation." Academic Press, New York. [34]

94. Cameron, I.L., and Greulich, R.C. (1963). Evidence for an essentially constant duration of DNA synthesis in renewing epithelia of the adult mouse. *J. Cell Biol.* **18**, 31–40. [46]

95. Cameron, I.L., and Prescott, D.M. (1961). Relations between cell growth and cell division. V. Cell and macronuclear volumes of *Tetrahymena pyriformis* HSM during the cell life cycle. *Exp. Cell Res.* **23**, 354–360. [14–15]

96. Charret, R., and André, J. (1968). La synthèse de l'ADN mitochondrial chez *Tetrahymena pyriformis.* Etude radioautographique quantitative au microscope électronique. *J. Cell Biol.* **39**, 369–381. [63]

97. Choe, B.-K., and Rose, N.R. (1974). Synthesis of DNA-binding protein in WI-38 cells stimulated to synthesize DNA by medium replacement. *Exp. Cell Res.* **83**, 261–270. [123]

98. Choe, B.-K., and Rose, N.R. (1974). Synthesis of DNA-binding proteins during the cell

cycle of WI-38 cells. *Exp. Cell Res.* **83**, 271–280. [123]

99. Church, K. (1967). Pattern of DNA replication in binucleate cells occurring in mouse embryo cell cultures. *Exp. Cell. Res.* **46**, 639–641. [61]

100. Churchill, J.R., and Studzinski, G.P. (1970). Thymidine as synchronizing agent. III. Persistence of cell cycle patterns of phosphatase activities and elevation of nuclease activity during inhibition of DNA synthesis. *J. Cell. Physiol.* **75**, 297–304. [128]

101. Cikes, M. (1970). Relationship between growth rate, cell volume, cell cycle kinetics, and antigenic properties of cultured murine lymphoma cells. *J. Natl. Cancer Inst.* **45**, 979–988. [43]

102. Clason, A.E., and Burdon, R.H. (1969). Synthesis of small nuclear ribonucleic acids of mammalian cells in relation to the cell cycle. *Nature (London)* **223**, 1063–1064. [125]

103. Clever, U. (1964). Actinomycin and puromycin: Effects on sequential gene activation by ecdysone. *Science* **146**, 794–795. [129]

104. Coffino, P., Gray, J.W., and Tomkins, G.M. (1975). Cyclic AMP, a nonessential regulator of the cell cycle. *Proc. Natl. Acad. Sci. U.S.A.* **72**, 878–882. [113]

105. Cohen, A.K., Rode, H.N., and Helleiner, C.W. (1972). The time of synthesis of satellite DNA in mouse cells (L cells). *Can. J. Biochem.* **50**, 229–231. [81]

106. Comings, D.E., and Okada, T.A. (1973). DNA replication and the nuclear membrane. *J. Mol. Biol.* **75**, 609–618. [68]

107. Comings, D.E., and Kakefuda, T. (1968). Initiation of deoxyribonucleic acid replication at the nuclear membrane in human cells. *J. Mol. Biol.* **33**, 225–229. [67]

108. Conrad, A.H. (1971). Thymidylate synthetase activity in cultured mammalian cells. *J. Biol. Chem.* **246**, 1318–1323. [56]

109. Cottrell, S.F., and Avers, C.J. (1970). Evidence of mitochondrial synchrony in synchronous cell cultures of yeast. *Biochem. Biophys. Res. Commun.* **38**, 973–980. [64]

110. Croce, C.M., and Koprowski, H. (1974). Positive control of transformed phenotype in hybrids between SV40-transformed and normal human cells. *Science* **184**, 1288–1289. [137]

111. Culotti, J., and Hartwell, L.H. (1971). Genetic control of the cell division cycle in yeast. III. Seven genes controlling nuclear division. *Exp. Cell Res.* **67**, 389–401. [131]

111a.Cummins, J.E., and Rusch, H.P. (1966). Limited DNA synthesis in the absence of protein synthesis in *Physarum polycephalum. J. Cell Biol.* **31**, 577–583. [50, 79]

112. Cunningham, D.D., and Pardee, A.B. (1969). Transport changes rapidly initiated by serum addition to "contact inhibited" 3T3 cells. *Proc. Natl. Acad. Sci. U.S.A.* **64**, 1049–1056. [125]

113. David, C.N., and Campbell, R.D. (1972). Cell cycle kinetics and development of *Hydra attenuata.* I. Epithelial cells. *J. Cell Sci.* **11**, 557–568. [51]

114. Davidson, D. (1966). The onset of mitosis and DNA synthesis in roots of germinating beans. *Am. J. Bot.* **53**, 491–495. [48–49]

115. Defendi, V., Ephrussi, B., Koprowski, H., and Yoshida, M.C. (1967). Properties of hybrids between polyoma-transformed and normal mouse cells. *Proc. Natl. Acad. Sci. U.S.A.* **57**, 299–305. [137]

116. Defendi, V., and Manson, L.A. (1963). Analysis of the life-cycle in mammalian cells. *Nature (London)* **198**, 359–361. [44]

117. de Morales, M.M., Blat, C., and Harel, L. (1974). Changes in the phosphorylation of non-histone chromosomal proteins in relationship to DNA and RNA synthesis in BHK_{21} C_{13} cells. *Exp. Cell Res.* **86**, 111–119. [123]

118. de Terra, N. (1967). Macronuclear DNA synthesis in Stentor: Regulation by a cytoplasmic initiator. *Proc. Natl. Acad. Sci. U.S.A.* **57**, 607–614. [64]

119. Dewey, W. C., Miller, H.H., and Nagasawa, H. (1973). Interactions between S and G1 cells. Effects on decay of synchrony. *Exp. Cell Res.* **77**, 73–78. [38]

120. Doida, Y., and Okada, S. (1967). Synchronization of L5178Y cells by successive

treatment with excess thymidine and colcemid. *Exp. Cell Res.* **48**, 540–548. [32]

121. Donachie, W.D., and Begg, K.J. (1970). Growth of the bacterial cell. *Nature (London)* **227**, 1220–1224. [40]

122. Donnelly, G.M., and Sisken, J.E. (1967). RNA and protein synthesis required for entry of cells into mitosis and during the mitotic cycle. *Exp. Cell Res.* **46**, 93–105. [89]

123. Duntze, W., Stötzler, D., Bücking-Throm, E., and Kalbitzer, S. (1973). Purification and partial characterization of α-factor, a mating-type specific inhibitor of cell reproduction from *Saccharomyces cerevisiae*. *Eur. J. Biochem.* **35**, 357–365. [5]

124. Eckhart, W., Dulbecco, R., and Burger, M.M. (1971). Temperature-dependent surface changes in cells infected or transformed by a thermosensitive mutant of polyoma virus. *Proc. Natl. Acad. Sci. U.S.A.* **68**, 283–286. [136]

125. Eker, P. (1974). Inhibition of growth and DNA synthesis in cell cultures by cyclic AMP. *J. Cell Sci.* **16**, 301–307. [112]

126. Elbrink, J., and Bihler, I. (1975). Membrane transport: Its relation to cellular metabolic rates. *Science* **188**, 1177–1184. [114]

127. Elgjo, K., Laerum, O.D., and Edgehill, W. (1972). Growth regulation in mouse epidermis. II. G_1-inhibitor present in the differentiating cell layer. *Virchows Arch. B* **10**, 229–236. [48]

128. Elgjo, K., Laerum, O.D., and Edgehill, W. (1971). Growth regulation in mouse epidermis. I. G_2-inhibitor present in the basal cell layer. *Virchows Arch. B* **8**, 277–283. [48]

129. Enger, M.D., and Tobey, R.A. (1972). Effects of isoleucine deficiency on nucleic acid and protein metabolism in cultured Chinese hamster cells. Continued ribonucleic acid and protein synthesis in the absence of deoxyribonucleic acid synthesis. *Biochemistry* **11**, 269–277. [55]

130. Enger, M.D., and Tobey, R.A. (1969). RNA synthesis in Chinese hamster cells. II. Increase in rate of RNA synthesis during G_1. *J. Cell Biol.* **42**, 308–315. [27, 124]

131. Ensminger, W.D., and Tamm, I. (1970). The step in cellular DNA synthesis blocked by Newcastle disease or mengovirus infection. *Virology* **40**, 152–165. [83]

132. Epifanova, O.I., Abuladze, M.K., and Zosimovskaya, A.I. (1975). Effects of low concentrations of actinomycin D on the initiation of DNA synthesis in rapidly proliferating and stimulated cell cultures. *Exp. Cell Res.* **92**, 23–30. [53]

133. Erlandson, R.A., and de Harven, E. (1971). The ultrastructure of synchronized HeLa cells. *J. Cell Sci.* **8**, 353–397. [68, 93]

134. Evans, L.S., and Van't Hof, J. (1973). Cell arrest in G2 in root meristems: A control factor from the cotyledons. *Exp. Cell Res.* **82**, 471–473. [88]

135. Evenson, D.P., and Prescott, D.M. (1970). Disruption of DNA synthesis in *Euplotes* by heat shock. *Exp. Cell Res.* **63**, 245–252. [71]

136. Everhart, L.P., Hauschka, P.V., and Prescott, D.M. (1973). Measurement of growth and rates of incorporation of radioactive precursors into macromolecules of cultured cells. *Methods Cell Biol.* **7**, 329–347. [31]

137. Everhart, L.P. (1972). Effects of deprivation of two essential amino acids on DNA synthesis in Chinese hamster cells. *Exp. Cell Res.* **74**, 311–318. [54, 83]

138. Everhart, L.P., and Prescott, D.M. (1972). Reversible arrest of Chinese hamster cells in G1 by partial deprivation of leucine. *Exp. Cell Res.* **75**, 170–174. [54]

139. Fabrikant, J.I., and Foster, B.R. (1969). Cell cycle of lymphocytes in mouse thymus. *Naturwissenschaften* **11**, 567–568. [43]

140. Fakan, S., Turner, G.N., Pagano, J.S., and Hancock, R. (1972). Sites of replication of chromosomal DNA in a eukaryotic cell. *Proc. Natl. Acad. Sci. U.S.A.* **69**, 2300–2305. [68, 72]

141. Fan, H., and Penman, S. (1971). Regulation of synthesis and processing of nucleolar components in metaphase-arrested cells. *J. Mol. Biol.* **59**, 27–42. [95]

142. Fan, H., and Penman, S. (1970). Regulation of protein synthesis in mammalian cells.

II. Inhibition of protein synthesis at the level of initiation during mitosis. *J. Mol. Biol.* **50**, 655–670. [93]

143. Fansler, B., and Loeb, L.A. (1972). Sea urchin nuclear DNA polymerase. IV. Reversible association of DNA polymerase with nuclei during the cell cycle. *Exp. Cell Res.* **75**, 433–441. [57–58]

144. Fansler, B., and Loeb, L.A. (1969). Sea urchin nuclear DNA polymerase. II. Changing localization during early development. *Exp. Cell Res.* **57**, 305–310. [51, 57–58]

145. Farber, J., Stein, G., and Baserga, R. (1972). The regulation of RNA synthesis during mitosis. *Biochem. Biophys. Res. Commun.* **47**, 790–797. [91]

146. Flamm, W.G., Bernheim, N.J., and Brubaker, P.E. (1971). Density gradient analysis of newly replicated DNA from synchronized mouse lymphoma cells. *Exp. Cell Res.* **64**, 97–104. [80]

147. Fouquet, H., and Braun, R. (1974). Differential RNA synthesis in the mitotic cycle of *Physarum polycephalum. FEBS Lett.* **38**, 184–186. [126]

148. Fournier, R.E., and Pardee, A.B. (1975). Cell cycle studies of mononucleate and cytochalasin-B-induced binucleate fibroblasts. *Proc. Natl. Acad. Sci. U.S.A.* **72**, 869–873. [41, 61]

149. Fox, T.O., Sheppard, J.R., and Burger, M.M. (1971). Cyclic membrane changes in animal cells: Transformed cells permanently display a surface architecture detected in normal cells only during mitosis. *Proc. Natl. Acad. Sci. U.S.A.* **68**, 244–247. [104, 136]

150. Fox, T.O., and Pardee, A.B. (1970). Animal cells: Noncorrelation of length of G1 phase with size after mitosis. *Science* **167**, 80–82. [40]

151. Franke, W.W., Deumling,B., Zentgraf, H., Falk, H., and Rae, P.M.M. (1973). Nuclear membranes from mammalian liver. IV. Characterization of membrane-attached DNA. *Exp. Cell Res.* **81**, 365–392. [72]

152. Frankfurt, O.S. (1968). Effect of hydrocortisone, adrenalin and actinomycin D on transition of cells to the DNA synthesis phase. *Exp. Cell Res.* **52**, 220–232. [53, 60]

153. Frazier, E.A.J. (1973). DNA synthesis following gross alterations of the nucleocytoplasmic ratio in the ciliate *Stentor coeruleus. Dev. Biol.* **34**, 77–92. [42]

154. Freed, J.J., and Schatz, S.A. (1969). Chromosome aberrations in cultured cells deprived of single essential amino acids. *Exp. Cell Res.* **55**, 393–409. [84]

155. Friedberg, S.H., and Davidson, D. (1970). Duration of S phase and cell cycle in diploid and tetraploid cells of mixoploid meristems. *Exp. Cell Res.* **61**, 216–218. [85]

156. Friedman, D.L. (1970). DNA polymerase from HeLa cell nuclei: Levels of activity during a synchronized cell cycle. *Biochem. Biophys. Res. Commun.* **39**, 100–109. [57]

157. Friedman, D.L., and Mueller, G.C. (1968). A nuclear system for DNA replication from synchronized HeLa cells. *Biochim. Biophys. Acta* **161**, 455–468. [66]

158. Froehlich, J.E., and Rachmeler, M. (1974). Inhibition of cell growth in the G_1 phase by adenosine 3′,5′-cyclic monophosphate. *J. Cell Biol.* **60**, 249–257. [113]

159. Fujiwara, Y. (1967). Role of RNA synthesis in DNA replication of synchronized populations of cultured mammalian cells. *J. Cell. Physiol.* **70**, 291–300. [85]

160. Galavazi, G., and Bootsma, D. (1966). Synchronization of mammalian cells *in vitro* by inhibition of the DNA synthesis. II. Population dynamics. *Exp. Cell Res.* **41**, 438–451. [32]

161. Gall, J.G. (1968). Differential synthesis of the genes for ribosomal RNA during amphibian oogenesis. *Proc. Natl. Acad. Sci. U.S.A.* **60**, 553–560. [82]

162. Gamow, E.I., and Prescott, D.M. (1970). The cell life cycle during early embryogenesis of the mouse. *Exp. Cell Res.* **59**, 117–123. [51]

163. Gateff, E., and Schneiderman, H.A. (1969). Neoplasms in mutant and cultured wild-type tissues of *Drosophila. Natl. Cancer Inst., Monogr.* **31**, 365–397. [136]

164. Gaulden, M.E. (1956). DNA synthesis and x-ray effects at different mitotic stages in grasshopper neuroblasts. *Genetics* 41, 645. [51, 87]
165. Gautschi, J.R. (1974). Effects of puromycin on DNA chain elongation in mammalian cells. *J. Mol. Biol.* 84, 223–229. [83]
166. Gautschi, J.R., and Kern, R.M. (1973). DNA replication in mammalian cells in the presence of cycloheximide. *Exp. Cell Res.* 80, 15–26. [83]
167. Gelbard, A.S., Perez, A.G., Kim, J.H., and Djordjevic, B. (1971). The effect of x-irradiation on thymidine kinase activity in synchronous populations of HeLa cells. *Radiat. Res.* 46, 334–342. [56]
168. Gelbard, A.S., Kim, J.H., and Perez, A.G. (1969). Fluctuations in deoxycytidine monophosphate deaminase activity during the cell cycle in synchronous populations of HeLa cells. *Biochim. Biophys. Acta* 182, 564–566. [57]
169. Gelfant, S. (1975). Temperature-induced cell proliferation in mouse ear epidermis *in vivo*. *Exp. Cell Res.* 90, 458–461. [88]
170. Gelfant, S. (1963). A new theory on the mechanism of cell division. *Symp. Int. Soc. Cell Biol.* 2, 229–259. [88]
171. Gerner, E.W., and Humphrey, R.M. (1973). The cell-cycle phase synthesis of non-histone proteins in mammalian cells. *Biochim. Biophys. Acta* 331, 117–127. [122]
172. Giacomoni, D., and Finkel, D. (1972). Time of duplication of ribosomal RNA cistrons in a cell line of *Potorous tridactylis* (rat kangaroo). *J. Mol. Biol.* 70, 725–728. [82]
173. Gilden, R.V., Carp, R.I., Taguchi, F., and Defendi, V. (1965). The nature and localization of the SV40-induced complement-fixing antigen. *Proc. Natl. Acad. Sci. U.S.A.* 53, 684–692. [95]
174. Giménez-Martín, G., González-Fernández, A., de la Torre, C., and Fernández-Gómez, M.E. (1971). Partial initiation of endomitosis by 3′-deoxyadenosine. *Chromosoma* 33, 361–371. [90]
175. Glinos, A.D., and Werrlein, R.J. (1972). Density dependent regulation of growth in suspension cultures of L-929 cells. *J. Cell. Physiol.* 79, 79–90. [44]
176. Goldstein, L. (1974). Movement of molecules between nucleus and cytoplasm. *In* "The Cell Nucleus" (H. Busch, ed.), Vol. 1, pp. 387–438. Academic Press, New York. [122]
177. Goldstein, L. (1970). On the question of protein synthesis by cell nuclei. *Adv. Cell Biol.* 1, 187–210. [122]
178. González-Fernández, A., Giménez-Martín, G., Díez, J.L., de la Torre, C., and López-Sáez, J.F. (1971). Interphase development and beginning of mitosis in the different nuclei of polynucleate homokaryotic cells. *Chromosoma* 36, 100–111. [61]
179. Gordon, M. (1959). The melanoma cell as an incompletely differentiated pigment cell. *In* "Pigment Cell Biology" (M. Gordon, ed.), pp. 215–239. Academic Press, New York. [136]
180. Graham, C.F. (1966). The effect of cell size and DNA content on the cellular regulation of DNA synthesis in haploid and diploid embryos. *Exp. Cell. Res.* 43, 13–19. [42, 85]
181. Graham, C.F., and Morgan, R.W. (1966). Changes in the cell cycle during early amphibian development. *Dev. Biol.* 14, 439–460. [47, 51]
182. Graham, C.F., Arms, K., and Gurdon, J.B. (1966). The induction of DNA synthesis by frog egg cytoplasm. *Dev. Biol.* 14, 349–381. [65]
183. Graham, F.L., van der Eb, A.J., and Heijneker, H.L. (1974). Size and location of the transforming region in human adenovirus type 5 DNA. *Nature (London)* 251, 687–691. [136]
184. Graham, J.M., Sumner, M.C.B., Curtis, D.H., and Pasternak, C.A. (1973). Sequence of events in plasma membrane assembly during the cell cycle. *Nature (London)* 246, 291–295. [103–104]

185. Granner, D., Chase, L.R., Aurbach, G.D., and Tomkins, G.M. (1968). Tyrosine aminotransferase: Enzyme induction independent of adenosine 3',5'-monophosphate. *Science* **162**, 1018–1020. [116]

186. Graves, J.A.M. (1972). DNA synthesis in heterokaryons formed by fusion of mammalian cells from different species. *Exp. Cell Res.* **72**, 393–403. [61]

187. Graves, J.A.M. (1972). Cell cycles and chromosome replication patterns in interspecific somatic hybrids. *Exp. Cell Res.* **73**, 81–94. [62]

188. Greaves, M.F., and Bauminger, S. (1972). Activation of T and B lymphocytes by insoluable phytomitogens. *Nature (London), New Biol.* **235**, 67–70. [115]

189. Green, H. (1974). Ribosome synthesis during preparation for division in the fibroblast. *In* "Control of Proliferation in Animal Cells," pp. 743–755. Cold Spring Harbor Lab., Cold Spring Harbor, New York. [123, 125–126]

190. Grimes, W.J., and Schroeder, J.L. (1973). Dibutyryl cyclic adenosine 3',5'-monophosphate, sugar transport, and regulatory control of cell division in normal and transformed cells. *J. Cell Biol.* **56**, 487–491. [112]

191. Grosset, L., and Odartchenko, N. (1975). Duration of mitosis and separate mitotic phases compared to nuclear DNA content in erythroblasts of four vertebrates. *Cell Tissue Kinet.* **8**, 91–96. [86]

192. Grove, G.L. (1974). A cytophotometric analysis of nuclear DNA contents of cultured human diploid cells in log and in plateau phases of growth. *Exp. Cell Res.* **87**, 386–387. [34, 44–45]

193. Groyon, R.M., and Kniazeff, A.J. (1967). Vaccinia virus infection of synchronized pig kidney cells. *J. Virol.* **1**, 1255–1264. [94]

194. Gurdon, J.B. (1967). On the origin and persistence of a cytoplasmic state inducing nuclear DNA synthesis in frogs' eggs. *Proc. Natl. Acad. Sci. U.S.A.* **58**, 545–552. [65]

195. Gurley, L.R., Walters, R.A., and Tobey, R.A. (1974). The metabolism of histone fractions. Phosphorylation and synthesis of histones in late G_1-arrest. *Arch. Biochem. Biophys.* **164**, 469–477. [120–121]

196. Gurley, L.R., Walters, R.A., and Tobey, R.A. (1973). Histone phosphorylation in late interphase and mitosis. *Biochem. Biophys. Res. Commun.* **50**, 744–750. [34, 120]

197. Gurley, L.R., Walters, R.A., and Tobey, R.A. (1973). The metabolism of histone fractions. VI. Differences in the phosphorylation of histone fractions during the cell cycle. *Arch. Biochem. Biophys.* **154**, 212–218. [120]

198. Gurley, L.R., Walters, R.A., and Tobey, R.A. (1972). The metabolism of histone fractions. IV. Synthesis of histones during the G_1-phase of the mammalian life cycle. *Arch. Biochem. Biophys.* **148**, 633–641. [120]

199. Guttes, E., and Guttes, S. (1969). Replication of nucleolus-associated DNA during "G_2 phase" in *Physarum polycephalum. J. Cell Biol.* **43**, 229–236. [82]

200. Guttes, E.W., Hanawalt, P.C., and Guttes, S. (1967). Mitochondrial DNA synthesis and the mitotic cycle in *Physarum polycephalum. Biochim. Biophys. Acta* **142**, 181–194. [63]

201. Guttes, S., and Guttes, E. (1968). Regulation of DNA replication in the nuclei of the slime mold *Physarum polycephalum.* Transplantation of nuclei by plasmodial coalescence. *J. Cell Biol.* **37**, 761–772. [64]

202. Hadden, J.W., Hadden, E.M., Haddox, M.K., and Goldberg, N.D. (1972). Guanosine 3',5'-cyclic monophosphate: A possible intracellular mediator of mitogenic influences in lymphocytes. *Proc. Natl. Acad. Sci. U.S.A.* **69**, 3024–3027. [114–115, 118]

203. Hahn, G.M., Stewart, J.R., Yang, S.-J., and Parker, V. (1968). Chinese hamster cell monolayer cultures. I. Changes in cell dynamics and modifications of the cell cycle with the period of growth. *Exp. Cell Res.* **49**, 285–292. [43–44]

204. Hale, A.H., Winkelhake, J.L., and Weber, M.J. (1975). Cell surface changes and rous sarcoma virus gene expression in synchronized cells *J. Cell Biol.* **64**, 398–407. [97, 103]

205. Hall, R.G., Jr. (1969). DNA synthesis in organ cultures of the hamster cheek pouch. *Exp. Cell Res.* **58**, 429–431. [48]

206. Hanania, N., Caneva, R., Tapiero, H., and Harel, J. (1975). Distribution of repetitious DNA in randomly growing and synchronized Chinese hamster cells. *Exp. Cell Res.* **90**, 79–86. [82]

207. Hanaoka, F., and Yamada, M. (1971). Localization of the replication point of mammalian cell DNA at the membrane. *Biochem. Biophys. Res. Commun.* **42**, 647–653. [68]

208. Hancock, R., and Weil, R. (1969). Biochemical evidence for induction by polyoma virus of replication of the chromosomes of mouse kidney cells. *Proc. Natl. Acad. Sci. U.S.A.* **63**, 1144–1150. [115]

209. Hand, R., and Tamm, I. (1973). DNA replication: Direction and rate of chain growth in mammalian cells. *J. Cell Biol.* **58**, 410–418. [75, 83]

210. Hand, R., and Tamm, I. (1972). Rate of DNA chain growth in mammalian cells infected with cytocidal RNA viruses. *Virology* **47**, 331–337. [75]

211. Hand, R., Ensminger, W.D., and Tamm, I. (1971). Cellular DNA replication in infections with cytocidal RNA viruses. *Virology* **44**, 527–536. [83]

212. Hanson, K.P., Ivanova, L.V., Nikitina, Z.S., Shutko, A.N., and Komar, V.E. (1970). Correlation of biosynthesis of mitochondrial and nuclear DNA in the process of regeneration of rat liver after partial hepatectomy. *Biokhimiya* **35**, 635–640. [63]

213. Harris, H., Miller, O.J., Klein, G., Worst, P., and Tachibana, T. (1969). Suppression of malignancy by cell fusion. *Nature (London)* **223**, 363–368. [137]

214. Hartwell, L.H. (1974). *Saccharomyces cerevisiae* cell cycle. *Bacteriol. Rev.* **38**, 164–198. [5, 131]

215. Hartwell, L.H. (1973). Synchronization of haploid yeast cell cycles, a prelude to conjugation. *Exp. Cell Res.* **76**, 111–117. [5]

216. Hartwell, L.H., Mortimer, R.K., Culotti, J., and Culotti, M. (1973). Genetic control of the cell division cycle in yeast. V. Genetic analysis of *cdc* mutants. *Genetics* **74**, 267–286. [53, 131]

217. Hartwell, L.H., Culotti, J., Pringle, J.R., and Reid, B.J. (1974). Genetic control of the cell division cycle in yeast. *Science* **183**, 46–51. [131–132]

218. Hartwell, L.H. (1971). Genetic control of the cell division cycle in yeast. II. Genes controlling DNA replication and its initiation. *J. Mol. Biol.* **59**, 183–194. [131]

219. Hartwell, L.H., Culotti, J., and Reid, B. (1970). Genetic control of the cell division cycle in yeast. I. Detection of mutants. *Proc. Natl. Acad. Sci. U.S.A.* **66**, 352–359. [131]

220. Hatch, F.T., and Mazrimas, J.A. (1970). Satellite DNA's in the kangaroo rat. *Biochim. Biophys. Acta* **224**, 291–294. [81]

221. Hatfield, J.M.R., and Walker, P.M.B. (1973). Satellite DNA replication in baby mouse kidney cells infected with polyoma virus. *Nature (London), New Biol.* **242**, 141–142. [81]

222. Hatzfeld, J., and Buttin, G. (1975). Temperature-sensitive cell cycle mutants: A Chinese hamster cell line with a reversible block in cytokinesis. *Cell* **5**, 123–129. [134]

223. Hauschka, P.V., Everhart, L.P., and Rubin, R.W. (1972). Alteration of nucleoside transport of Chinese hamster cells by dibutyryl adenosine 3':5'-cyclic monophosphate. *Proc. Natl. Acad. Sci. U.S.A.* **69**, 3542–3546. [114]

224. Heidrick, M.L., and Ryan, W.L. (1971). Metabolism of 3',5'-cyclic AMP by strain L cells. *Biochim. Biophys. Acta* **237**, 301–309. [114]

225. Hereford, L.M., and Hartwell, L.H. (1973). Role of protein synthesis in the replication of yeast DNA. *Nature (London), New Biol.* **244**, 129–131. [52–53, 84]

226. Hershey, H.V., Stieber, J.F., and Mueller, G.C. (1973). DNA synthesis in isolated HeLa nuclei. A system for continuation of replication *in vivo. Eur. J. Biochem.* **34**, 383–394. [66]

227. Hertwig, R. (1908). Neue probl. d. zellenlehre. *Arch. Zellforsch* **1**, 1–32. [7]

228. Highfield, D.P., and Dewey, W.C. (1972). Inhibition of DNA synthesis in synchronized Chinese hamster cells treated in G1 or early S phase with cycloheximide or puromycin. *Exp. Cell Res.* **75**, 314–320. [52–53]

229. Hill, R.N., and Yunis, J.J. (1967). Mammalian x-chromosomes: Change in patterns of DNA replication during embryogenesis. *Science* **155**, 1120–1121. [81]

230. Hinegardner, R.T., Rao, B., and Feldman, D.E. (1964). The DNA synthetic period during early development of the sea urchin egg. *Exp. Cell Res.* **36**, 53–61. [51, 58]

231. Hodge, L.D., Robbins, E., and Scharff, M.D. (1969). Persistence of messenger RNA through mitosis in HeLa cells. *J. Cell Biol.* **40**, 497–507. [93]

232. Holley, R.W., and Hiernan, J.A. (1974). Control of the initiation of DNA synthesis in 3T3 cells: Serum factors. *Proc. Natl. Acad. Sci. U.S.A.* **71**, 2908–2911. [4]

233. Holley, R.W., and Kiernan, J.A. (1974). Control of the initiation of DNA synthesis in 3T3 cells: Low-molecular-weight nutrients. *Proc. Natl. Acad. Sci. U.S.A.* **71**, 2942–2945. [4]

233a. Holley, R.W., Baldwin, J.H., and Kiernan, J.A. (1974). Control of growth of a tumor cell by linoleic acid. *Proc. Natl. Acad. Sci. U.S.A.* **71**, 3976–3976–3978. [4]

234. Hori, T., and Lark, K.G. (1973). Effect of puromycin on DNA replication in Chinese hamster cells. *J. Mol. Biol.* **77**, 391–404. [76, 83]

235. Hoshino, K., Matsuzawa, T., and Murakami, U. (1973). Characteristics of the cell cycle of matrix cells in the mouse embryo during histogenesis of telencephalon. *Exp. Cell Res.* **77**, 89–94. [47]

236. Houck, J.C., and Hennings, H. (1973). Chalones: Specific endogenous mitotic inhibitors. *FEBS Lett.* **32**, 1–8. [47–48, 118]

237. Howard, A., and Pelc, S.R. (1953). Synthesis of deoxyribonucleic acid in normal and irradiated cells and its relation to chromosome breakage. *Heredity, Suppl.* **6**, 261–273. [3, 29]

238. Howard, D.K., Hay, J., Melvin, W.T., and Durham, J.P. (1974). Changes in DNA and RNA synthesis and associated enzyme activities after the stimulation of serum-depleted BHK21/C13 cells by the addition of serum. *Exp. Cell Res.* **86**, 31–42. [56–57]

239. Howell, S.H., and Naliboff, J.A. (1973). Conditional mutants in *Chlamydomonas reinhardtii* blocked in the vegetative cell cycle. I. An analysis of cell cycle block points. *J. Cell Biol.* **57**, 760–772. [133]

240. Huberman, J.A., Tsai, A., and Deich, R.A. (1973). DNA replication sites within nuclei of mammalian cells. *Nature (London)* **241**, 32–36. [68, 75]

241. Huberman, J.A., and Riggs, A.D. (1968). On the mechanism of DNA replication in mammalian chromosomes. *J. Mol. Biol.* **32**, 327–341. [74–75]

242. Jakob, K.M., and Bovey, F. (1969). Early nucleic acid and protein syntheses and mitoses in the primary root tips of germinating *Vicia faba. Exp. Cell Res.* **54**, 118–126. [48]

243. James, T.W. (1966). Cell synchrony, a prologue to discovery. *In* "Cell Synchrony: Studies in Biosynthetic Regulation" (I.L. Cameron and G.M. Padilla, eds.), pp. 1–13. Academic Press, New York. [19, 21]

244. Jiminez de Asua, L., Rozengurt, E., and Dulbecco, R. (1974). Kinetics of early

changes in phosphate and uridine transport and cyclic AMP levels stimulated by serum in density-inhibited 3T3 cells. *Proc. Natl. Acad. Sci. U.S.A.* **71**, 96–98. [114]

245. Johnson, E.M., and Hadden, J.W. (1975). Phosphorylation of lymphocyte nuclear acidic proteins: Regulation by cyclic nucleotides. *Science* **187**, 1198–1200. [118]

246. Johnson, R.A., and Schmidt, R.R. (1966). Enzymic control of nucleic acid synthesis during synchronous growth of *Chlorella pyrenoidosa*. I. Deoxythymidine monophosphate kinase. *Biochim. Biophys. Acta* **129**, 140–144. [57]

247. Johnson, R.T., and Rao, P.N. (1970). Mammalian cell fusion: Induction of premature chromosome condensation in interphase nuclei. *Nature (London)* **226**, 717–722. [87]

248. Johnson, R.T., Rao, P.N., and Hughes, D.S. (1970). Mammalian cell fusion. III. A HeLa cell inducer of premature chromosome condensation active in cells from a variety of animal species. *J. Cell. Physiol.* **76**, 151–158. [88]

249. Johnson, R.T., and Harris, H. (1969). DNA synthesis and mitosis in fused cells. II. HeLa-chick erythrocyte heterokaryons. *J. Cell Sci.* **5**, 625–644. [61, 65]

250. Kauffmann, S.L. (1968). Lengthening of the generation cycle during embryonic differentiation of the mouse neural tube. *Exp. Cell Res.* **49**, 420–424. [51]

251. Kay, R.R., Haines, M.E., and Johnston, I.R. (1971). Late replication of the DNA associated with the nuclear membrane. *FEBS Lett.* **16**, 233–236. [68]

252. Killander, D., and Zetterberg, A. (1965). Quantitative cytochemical studies on interphase growth. I. Determination of DNA, RNA, and mass content of age determined mouse fibroblasts *in vitro* and of intercellular variation in generation time. *Exp. Cell Res.* **38**, 272–284. [16, 39–40]

253. Killander, D., and Zetterberg, A. (1965). A quantitative cytochemical investigation of the relationship between cell mass and initiation of DNA synthesis in mouse fibroblasts *in vitro*. *Exp. Cell Res.* **40**, 12–20. [7, 39–40]

254. Kim, J.H., Kim, S.H., and Eidinoff, M.L. (1965). Cell viability and nucleic acid metabolism after exposure of HeLa cells to excess thymidine and deoxyadenosine. *Biochem. Pharmacol.* **14**, 1821–1829. [32]

255. Kimball, R.F., Perdue, S.W., Chu, E.H.Y., and Ortiz, J.R. (1971). Microphotometric and autoradiographic studies on the cell cycle and cell size during growth and decline of Chinese hamster cell cultures. *Exp. Cell Res.* **66**, 17–32. [42–43]

256. Kimball, R.F., and Prescott, D.M. (1962). Deoxyribonucleic acid synthesis and distribution during growth and amitosis of the macronucleus of *Euplotes*. *J. Protozool.* **9**, 88–92. [49, 61–62]

257. Kimball, R.F., and Barka, T. (1959). Quantitative cytochemical studies on *Paramecium aurelia*. II. Feulgen microspectrophotometry of the macronucleus during exponential growth. *Exp. Cell Res.* **17**, 173–182. [62]

258. Kinsey, J.D. (1967). X-chromosome replication in early rabbit embryos. *Genetics* **55**, 337–343. [81]

259. Kishimoto, S., and Lieberman, I. (1964). Synthesis of RNA and protein required for the mitosis of mammalian cells. *Exp. Cell Res.* **36**, 92–101. [52, 89]

260. Klein, G., Bregula, U., Wiener, F., and Harris, H. (1971). The analysis of maligancy by cell fusion. I. Hybrids between tumour cells and L cell derivatives. *J. Cell Sci.* **8**, 659–672. [137]

261. Klevecz, R.R. (1969). Temporal order in mammalian cells. I. The periodic synthesis of lactate dehydrogenase in the cell cycle. *J. Cell Biol.* **43**, 207–219. [128]

262. Klevecz, R.R., and Ruddle, F.H. (1968). Cyclic changes in synchronized mammalian cell cultures. *Science* **159**, 634–636. [128]

263. Klevecz, R.R., and Stubblefield, E. (1967). RNA synthesis in relation to DNA replication in synchronized Chinese hamster cell cultures. *J. Exp. Zool.* **165**, 259–268. [124]

264. Koburg, E., and Maurer, W. (1962). Autoradiographische untersuchung mit(^{3}H)

thymidin über die dauer der deoxyribonukleinsäure-synthese und ihren zeitlichen verlauf bei den darmepithelien und anderen zelltypen der maus. *Biochim. Biophys. Acta* **61**, 229–242. [88]

265. Koch, J., and Stokstad, E.L.R. (1967). Incorporation of ³H-thymidine into nuclear and mitochondrial DNA in synchronized mammalian cells. *Eur. J. Biochem.* **3**, 1. [63]

266. Kolodny, G.M., and Gross, P.R. (1969). Changes in patterns of protein synthesis during the mammalian cell cycle. *Exp. Cell Res.* **56**, 117–121. [90]

267. Kosswig, C. (1964). Problems of polymorphism in fishes. *Copeia* **1964**, 65–75. [136]

268. Kraemer, P.M., and Tobey, R.A. (1972). Cell-cycle dependent desquamation of heparan sulfate from the cell surface. *J. Cell Biol.* **55**, 713–717. [106]

269. Kram, R., Mamont, P., and Tomkins, G.M. (1973). Pleiotypic control by adenosine 3':5'-cyclic monophosphate: A model for growth control in animal cells. *Proc. Natl. Acad. Sci. U.S.A.* **70**, 1432–1436. [113–114]

270. Kriegstein, H.J., and Hogness, D.S. (1974). Mechanism of DNA replication in *Drosophila* chromosomes: Structure of replication forks and evidence for bidirectionality. *Proc. Natl. Acad. Sci. U.S.A.* **71**, 135–139. [77–78]

271. Kubitschek, H.E. (1971). The distribution of cell generation times. *Cell Tissue Kinet.* **4**, 113–122. [24]

272. Kubitschek, H.E. (1970). Evidence for the generality of linear cell growth. *J. Theor. Biol.* **28**, 15–29. [8]

273. Kuehn, G.D. (1972). Cell cycle variation in cyclic adenosine 3',5'-monophosphate-dependent inhibition of a protein kinase from *Physarum polycephalum*. *Biochem. Biophys. Res. Commun.* **49**, 414–419. [118]

274. Kuempel. P.L. (1970). Bacterial chromosome replication. *Adv. Cell Biol.* **1**, 3–56. [67, 73]

275. Kumar, K.V., and Friedman, D.L. (1972). Initiation of DNA synthesis in HeLa cell-free system. *Nature (London), New Biol.* **239**, 74–76. [66]

276. Kumon, A., Nishiyama, K., Yamamura, H., and Nishizuka, Y. (1972). Multiplicity of adenosine 3',5'-monophosphate-dependent protein kinases from rat liver and mode of action of nucleoside 3',5'-monophosphate. *J. Biol. Chem.* **247**, 3726–3735. [118]

277. Kurnick, N.G., and Herskowitz, I.H. (1952). The estimation of polyteny in *Drosophila* salivary gland nuclei based on determination of deoxyribonucleic acid content. *J. Cell. Comp. Physiol.* **39**, 281–299. [130]

278. Laird, C.D. (1971). Chromatid structure: Relationship between DNA content and nucleotide sequence diversity. *Chromosoma* **32**, 378–406. [74, 130]

279. Lajtha, L.G. (1963). On the concept of the cell cycle. *J. Cell. Comp. Physiol.* **62**, 143–145. [6]

280. Lake, R.S. (1973). F1-histone phosphorylation in metaphase chromosomes of cultured Chinese hamster cells. *Nature (London), New Biol.* **242**, 145–146. [120–121]

281. Lake, R.S., Goidl, J.A., and Salzman, N.P. (1972). F1-histone modification at metaphase in Chinese hamster cells. *Exp. Cell Res.* **73**, 113–121. [120]

282. Lala, P.K., and Patt, H.M. (1966). Cytokinetic analysis of tumor growth. *Proc. Natl. Acad. Sci. U.S.A.* **56**, 1735–1742. [43, 51]

283. Lark, K.G., Consigli, R., and Toliver, A. (1971). DNA replication in Chinese hamster cells: Evidence for a single replication fork per replicon. *J. Mol. Biol.* **58**, 873–875. [75]

284. Lawrence, P.A. (1968). Mitosis and the cell cycle in the metamorphic moult of the milkweed bug, *Oncopeltus fasciatus*. A radioautographic study. *J. Cell Sci.* **3**, 391–404. [87]

285. Lehmann, A.R., and Ormerod, M.G. (1970). The replication of DNA in murine lymphoma cells (L5178Y). I. Rate of replication. *Biochim. Biophys. Acta* **204**, 128–143. [75]

286. Levisohn, S.R., and Thompson, E.B. (1973). Contact inhibition and gene expression in HTC/L cell hybrid lines. *J. Cell. Physiol.* **81**, 225–232. [138]

287. Ley, K.D., and Murphy, M.M. (1973). Synchronization of mitochondrial DNA syn-

thesis in Chinese hamster cells (line CHO) deprived of isoleucine. *J. Cell Biol.* **58**, 340–345. [63]

288. Ley, K.D., and Tobey, R.A. (1970). Regulation of initiation of DNA synthesis in Chinese hamster cells. II. Induction of DNA synthesis and cell division by isoleucine and glutamine in G_1-arrested cells in suspension culture. *J. Cell Biol.* **47**, 453–459. [44]

289. Lin, M.S., and Davidson, R.L. (1975). Replication of human chromosomes in human–mouse hybrids: Evidence that the timing of DNA synthesis is determined independently in each human chromosome. *Somatic Cell Genet.* **1**, 111–122. [79–80]

290. Liskay, R.M. (1974). A mammalian somatic "cell cycle" mutant defective in G_1. *J. Cell. Physiol.* **84**, 49–56. [134]

291. Littlefield, J.W., McGovern, A.P., and Margeson, K.B. (1963). Changes in the distribution of polymerase activity during DNA synthesis in mouse fibroblasts. *Proc. Natl. Acad. Sci. U.S.A.* **49**, 102–108. [57]

292. Löbbecke, E.-A., Schultze, B., and Maurer, W. (1969). Variabilitat der generationszeit bei fetalen zellarten der ratte. Autoradiographische untersuchungen nach dauerinfusion mit ³H-thymidin. *Exp. Cell Res.* **55**, 176–184. [47]

293. Loeb, L.A., Ewald, J.L., and Agarwal, S.S. (1970). DNA polymerase and DNA replication during lymphocyte transformation. *Cancer Res.* **30**, 2514–2520. [55, 57]

294. Loeb, L.A., and Fansler, B. (1970). Intracellular migration of DNA polymerase in early developing sea urchin embryos. *Biochim. Biophys. Acta* **217**, 50–55. [58]

295. Loeb, L.A., Fansler, B., Williams, R., and Mazia, D. (1969). Sea urchin nuclear DNA polymerase. I. Localization in nuclei during rapid DNA synthesis. *Exp. Cell Res.* **57**, 298–304. [57]

296. Louie, A.J., and Dixon, G.H. (1973). Kinetics of phosphorylation and dephosphorylation of testis histones and their possible role in determining chromosomal structure. *Nature (London), New Biol.* **243**, 164–168. [120]

297. Lövlie, A. (1963). Growth in mass and respiration rate during the cell cycle of *Tetrahymena pyriformis*. *C. R. Trav. Lab. Carlsberg* **33**, 377–413. [11]

298. Lowdon, M., and Vitols, E. (1973). Ribonucleotide reductase activity during the cell cycle of *Saccharomyces cerevisiae*. *Arch. Biochem. Biophys.* **158**, 177–184. [57]

299. Madreiter, H., Kaden, P., and Mittermayer, C. (1971). DNA polymerase, triphosphatase and deoxyribonuclease in a system of synchronized L cells. *Eur. J. Biochem.* **18**, 369–375. [57, 128]

300. Makman, M.H., and Klein, M.I. (1972). Expression of adenylate cyclase, catecholamine receptor, and cyclic adenosine monophosphate-dependent protein kinase in synchronized culture of Chang's liver cells. *Proc. Natl. Acad. Sci. U.S.A.* **69**, 456–458. [109]

301. Marin, G., and Colletta, G. (1974). Patterns of late chromosomal DNA replication in unbalanced Chinese hamster–mouse somatic cell hybrids. *Exp. Cell Res.* **89**, 368–376. [79]

302. Marks, D.B., Paik, W.K., and Borun, T.W. (1973). The relationship of histone phosphorylation to deoxyribonucleic acid replication and mitosis during the HeLa S-3 cell cycle. *J. Biol. Chem.* **248**, 5660–5667. [120]

303. Martin, D.W., Jr., and Tomkins, G.M. (1970). The appearance and disappearance of the post-transcriptional repressor of tyrosine amino-transferase synthesis during the HTC cell cycle. *Proc. Natl. Acad. Sci. U.S.A.* **65**, 1064–1068. [128]

304. Martin, R.G., and Chou, J.Y. (1975). Simian virus 40 functions required for the establishment and maintenance of malignant transformation. *J. Virol.* **15**, 599–612. [136]

305. Matsui, S.-I., Yoshida, H., Weinfeld, H., and Sandberg, A.A. (1972). Induction of prophase in interphase nuclei by fusion with metaphase cells. *J. Cell Biol.* **54**, 120–132. [87]

306. Mazia, D., and Dan, K. (1952). The isolation and biochemical characterization of the mitotic apparatus of dividing cells. *Proc. Natl. Acad. Sci. U.S.A.* **38**, 826–838. [20]

307. Mazia, D. (1961). Mitosis and the physiology of cell division. *In* "The Cell" (J. Brachet and A.E. Mirsk, eds.), Vol. 3, pp. 77–412. Academic Press, New York. [8]

308. Mazrimas, J.A., and Hatch, F.T. (1972). A possible relationship between satellite DNA and the evolution of kangaroo rat species (genus *Dipodomys*). *Nature (London), New Biol.* **240**, 102–105. [76]

309. McDonald, B. (1962). Synthesis of deoxyribonucleic acid by micro- and macronuclei of *Tetrahymena pyriformis*. *J. Cell Biol.* **13**, 193–203. [49, 62]

310. McDonald, B.B. (1958). Quantitative aspects of deoxyribose nucleic acid (DNA) metabolism in an amicronucleate strain of *Tetrahymena*. *Biol. Bull. (Woods Hole, Mass.)* **114**, 71. [62]

311. Mendelsohn, M.L., Dohan, F.C., Jr., and Moore, H.A., Jr. (1960). Autoradiographic analysis of cell proliferation in spontaneous breast cancer of C3H mouse. I. Typical cell cycle and timing of DNA synthesis. *J. Natl. Cancer Inst.* **25**, 477–484. [44]

312. Meyer, R.R., and Ris, H. (1966). Incorporation of tritiated thymidine and tritiated deoxyadenosine into mitochondrial DNA of chick fibroblasts. *J. Cell Biol.* **31**, 76A. [63]

313. Meyn, R.E., Hewitt, R.R., and Humphrey, R.M. (1973). Evaluation of S phase synchronization by analysis of DNA replication in 5-bromo-deoxyuridine. *Exp. Cell Res.* **82**, 137–142. [33, 39]

314. Millis, A.J.T., Forrest, G.A., and Pious, D.A. (1974). Cyclic AMP dependent regulation of mitosis in human lymphoid cells. *Exp. Cell Res.* **83**, 335–343. [109–110]

315. Millis, A.J.T., Forrest, G., and Pious, D.A. (1972). Cyclic AMP in cultured human lymphoid cells: Relationship to mitosis. *Biochem. Biophys. Res. Commun.* **49**, 1645–1649. [107, 116]

316. Mitchell, A.D., and Hoogenraad, N.J. (1975). *De novo* pyrimidine nucleotide biosynthesis in synchronized rat hepatoma (HTC) cells and mouse embryo fibroblast (3T3) cells. *Exp. Cell Res.* **93**, 105–110. [128]

317. Mitchison, J.M. (1973). The cell cycle of a eukaryote. *Symp. Soc. Gen. Microbiol.* **23**, 189–208. [127]

318. Mitchison, J.M. (1971). "The Biology of the Cell Cycle." Cambridge Univ. Press, London and New York. [8–9, 21, 34, 127, 129]

319. Mitchison, J.M. (1969). Enzyme synthesis in synchronous cultures. *Science* **165**, 657–663. [127]

320. Mitchison, J.M., and Creanor, J. (1969). Linear synthesis of sucrase and phosphatases during the cell cycle of *Schizosaccharomyces pombe*. *J. Cell Sci.* **5**, 373–391. [127]

321. Mitchison, J.M., and Vincent, W.S. (1965). Preparation of synchronous cell cultures by sedimentation. *Nature (London)* **205**, 987–989. [21]

322. Mitchison, J.M., and Cummins, J.E. (1964). Changes in the acid-soluble pool during the cell cycle of *Schizosaccharomyces pombe*. *Exp. Cell Res.* **35**, 394–401. [10]

323. Mitchison, J.M., and Wilbur, K.M. (1962). The incorporation of protein and carbohydrate precursors during the cell cycle of fission yeast. *Exp. Cell Res.* **26**, 144–157. [10]

324. Mitchison, J.M., and Lark, K.G. (1962). Incorporation of ^{3}H-adenine into RNA during the cell cycle of *Schizosaccharomyces pombe*. *Exp. Cell Res.* **28**, 452–455. [10]

325. Miyashita, K., and Kakunaga, T. (1975). Isolation of heat- and cold-sensitive mutants of Chinese hamster lung cells affected in their ability to express the transformed state. *Cell* **5**, 131–138. [139]

326. Mizuno, N.S., Stoops, C.E., and Sinha, A.A. (1971). DNA synthesis associated with the inner membrane of the nuclear envelope. *Nature (London), New Biol.* **229**, 22–24. [68]

327. Moens, W., Vokaer, A., and Kram, R. (1975). Cyclic AMP and cyclic GMP concentrations in serum- and density-restricted fibroblast cultures. *Proc. Natl. Acad. Sci. U.S.A.*

72, 1063–1067. [113–114]

328. Monahan, T.M., Fritz, R.R., and Abell, C.W. (1973). Levels of cyclic AMP in murine L5178Y lymphoblasts grown in different concentrations of serum. *Biochem. Biophys. Res. Commun.* **55**, 642–646. [115]

329. Moore, E.C., and Hurlbert, R.B. (1966). Regulation of mammalian deoxyribonucleotide biosynthesis by nucleotides as activators and inhibitors. *J. Biol. Chem.* **241**, 4802–4809. [57]

330. Morris, N.R., Reichard, P., and Fischer, G.A. (1963). Studies concerning the inhibition of cellular reproduction by deoxyribonucleosides. II. Inhibition of the synthesis of deoxycytidine by thymidine, deoxyadenosine, and deoxyguanosine. *Biochim. Biophys. Acta* **68**, 93–99. [32]

331. Morris, N.R., and Fischer, G.A. (1963). Studies concerning the inhibition of cellular reproduction by deoxyribonucleosides. I. Inhibition of the synthesis of deoxycytidine by a phosphorylated derivative of thymidine. *Biochim. Biophys. Acta* **68**, 84–92. [32]

332. Mowat, D., Pearlman, R.E., and Engberg, J. (1974). DNA synthesis following refeeding of starved *Tetrahymena pyriformis* GL. Starved cells are arrested in G_1. *Exp. Cell Res.* **84**, 282–286. [45]

333. Mueller, G.C., and Kajiwara, K. (1966). Early- and late-replicating deoxyribonucleic acid complexes in HeLa nuclei. *Biochim. Biophys. Acta* **114**, 108–115. [79]

334. Mueller, G.C., and Kajiwara, K. (1966). Actinomycin D and p-fluorophenylalanine, inhibitors of nuclear replication in HeLa cells. *Biochim. Biophys. Acta* **119**, 557–565. [84]

335. Mukherjee, B.B., and Ghosal, S.K. (1969). Replicative differentiation of mammalian sex-chromosomes during spermatogenesis. *Exp. Cell Res.* **54**, 101–106. [81]

336. Mulder, M.P., van Duijn, P., and Gloor, H.J. (1968). The replicative organization of DNA in polytene chromosomes of *D. hydei. Genetica* **39**, 385–428. [74, 82]

337. Muldoon, J.J., Evans, T.E., Nygaard, O.F., and Evans, H.H. (1971). Control of DNA replication by protein synthesis at defined times during the S period in *Physarum polycephalum. Biochim. Biophys. Acta* **247**, 310–321. [79, 84]

338. Murphree, S., Stubblefield, E., and Moore, E.C. (1969). Synchronized mammalian cell cultures. III. Variation of ribonucleotide reductase activity during the replication cycle of Chinese hamster fibroblasts. *Exp. Cell Res.* **58**, 118–124. [57]

339. Newlon, C.S., Petes, T.D., Hereford, L.M., and Fangman, W.L. (1974). Replication of yeast chromosomal DNA. *Nature (London)* **247**, 32–35. [84]

340. Newlon, C.S., Sonenshein, G.E., and Holt, C.E. (1973). Time of synthesis of genes for ribosomal ribonucleic acid in *Physarum. Biochemistry* **12**, 2338–2345. [82]

341. Nexφ, B.A. (1975). Ribo- and deoxyribonucleotide triphosphate pools in synchronized populations of *Tetrahymena pyriformis. Biochim. Biophys. Acta* **378**, 12–17. [57]

342. Neyfakh, A.A., Abramova, N.B., and Bagrova, A.M. (1971). Migration of newly synthesized RNA during mitosis. II. Chinese hamster fibroblasts. *Exp. Cell Res.* **65**, 345–352. [95]

343. Neyfakh, A.A., and Kostomarova, A.A. (1971). Migration of newly synthesized RNA during mitosis. I. Embryonic cells of the Loach (*Misgurnus fossilis* L.). *Exp. Cell Res.* **65**, 340–344. [95]

344. Nicklas, R.B., and Jaqua, R.A. (1965). X chromosome DNA replication: Developmental shift from synchrony to asynchrony. *Science* **147**, 1041–1043. [81]

345. Nilausen, K., and Green, H. (1965). Reversible arrest of growth in G1 of an established fibroblast line (3T3). *Exp. Cell Res.* **40**, 166–168. [34, 44]

346. Noonan, K.D., and Burger, M.M. (1973). Induction of 3T3 cell division at the monolayer stage. Early changes in macromolecular processes. *Exp. Cell Res.* **80**, 405–414. [46]

347. Noonan, K.D., Levine, A.J., and Burger, M.M. (1973). Cell cycle-dependent changes in the surface membrane as detected with [^{3}H]Concanavalin A. *J. Cell Biol.* **58**, 491–497. [104–105]

348. Nose, K., and Katsuta, H. (1975). Arrest of cultured rat liver cells in G_2 phase by the treatment with dibutyryl cAMP. *Biochem. Biophys. Res. Commun.* **64**, 983–988. [88, 116]

349. Nur, U. (1966). Nonreplication of heterochromatic chromosomes in a mealy bug, *Planococcus citri* (Coccoidea: Homoptera). *Chromosoma* **19**, 439–448. [82]

350. Nygaard, O., Guttes, E., and Rusch, H.P. (1960). Nucleic acid metabolism in a slime mold with synchronous mitosis. *Biochim. Biophys. Acta* **38**, 298–306. [49, 60]

351. Oey, J., Vogel, A., and Pollack, R. (1974). Intracellular cyclic AMP concentration responds specifically to growth regulation by serum. *Proc. Natl. Acad. Sci. U.S.A.* **71**, 694–698. [113–114]

352. Oler, A., Iannaccone, P.M., and Gordon, G.B. (1973). Suppression of growth of L cells in suspension culture by dibutyryl adenosine 3',5'-monophosphate. *In Vitro* **9**, 35–38. [112]

353. Oliver, D., Balhorn, R., Granner, D., and Chalkley, R. (1972). Molecular nature of F_1 histone phosphorylation in cultured hepatoma cells. *Biochemistry* **11**, 3921–3925. [120]

354. Oppenheim, A., and Wahrman, J. (1973). DNA-membrane association during the mitotic cycle of *Physarum polycephalum. Exp. Cell Res.* **79**, 287–294. [72]

355. Ord, M.J. (1969). Control of DNA synthesis in *Amoeba proteus. Nature (London)* **221**, 964–966. [64]

356. Ord, M.J. (1968). The synthesis of DNA through the cell cycle of *Amoeba proteus. J. Cell Sci.* **3**, 483–491. [49]

357. Osborn, M., and Weber, K. (1975). Simian virus 40 gene *A* function and maintenance of transformation. *J. Virol.* **15**, 636–644. [136]

358. Otten, J., Bader, J., Johnson, G.S., and Pastan, I. (1972). A mutation in a rous sarcoma virus gene that controls adenosine 3',5'-monophosphate levels and transformation. *J. Biol. Chem.* **247**, 1632–1633. [113–114, 117]

359. Otten, J., Johnson, G.S., and Pastan, I. (1971). Cyclic AMP levels in fibroblasts: Relationships to growth rate and contact inhibition of growth. *Biochem. Biophys. Res. Commun.* **44**, 1192–1198. [114, 116]

360. Øye, I., and Sutherland, E.W. (1966). The effect of epinephrine and other agents on adenyl cyclase in the cell membrane of avian erythrocytes. *Biochim. Biophys. Acta* **127**, 347–354. [117]

361. Pages, J., Manteuil, S., Stehelin, D., Fiszman, M., Marx, M., and Girard, M. (1973). Relationship between replication of simian virus 40 DNA and specific events of the host cell cycle. *J. Virol.* **12**, 99–107. [115–116]

362. Pagoulatos, G.N., and Darnell, J.E. (1970). A comparison of the heterogeneous nuclear RNA of HeLa cells in different periods of the cell growth cycle. *J. Cell Biol.* **44**, 476–483. [126]

363. Painter, R.B., and Schaefer, A.W. (1971). Variation in the rate of DNA chain growth through the S phase in HeLa cells. *J. Mol. Biol.* **58**, 289–295. [75]

364. Painter, R.B., Jermany, D.A., and Rasmussen, R.E. (1966). A method to determine the number of DNA replicating units in cultured mammalian cells. *J. Mol. Biol.* **17**, 47–55. [74]

365. Panyim, S., and Chalkley, R. (1969). A new histone found only in mammalian tissues with little cell division. *Biochem. Biophys. Res. Commun.* **37**, 1042–1049. [120]

366. Papahadjopoulos, D., Poste, G., and Mayhew, E. (1974). Cellular uptake of cyclic AMP captured within phospholipid vesicles and effect on cell-growth behavior. *Biochim. Biophys. Acta* **363**, 404–418. [112]

367. Pardee, A.B. (1974). A restriction point for control of normal animal cell proliferation. *Proc. Natl. Acad. Sci. U.S.A.* **71**, 1286–1290. [4, 46]

368. Parsons, J.A. (1965). Mitochondrial incorporation of tritiated thymidine in *Tetrahymena pyriformis. J. Cell Biol.* **25**, 641–646. [63]

369. Pearson, G.D., and Hanawalt, P.C. (1971). Isolation of DNA replication complexes from uninfected and adenovirus-infected HeLa cells. *J. Mol. Biol.* **62**, 65–80. [68]

370. Pederson, T., and Gelfant, S. (1970). G2-population cells in mouse kidney and duodenum and their behavior during the cell division cycle. *Exp. Cell Res.* **59**, 32–36. [49, 88, 108]

371. Pederson, T., and Robbins, E. (1970). Absence of translational control of histone synthesis during the HeLa cell life cycle. *J. Cell Biol.* **45**, 509–513. [119]

372. Pelling, C. (1966). A replicative and synthetic chromosomal unit—the modern concept of the chromomere. *Proc. R. Soc. London, Ser. B* **164**, 279. [74]

373. Petersen, D.F., Anderson, E.C., and Tobey, R.A. (1968). Mitotic cells as a source of synchronized cultures. *Methods Cell Physiol.* **3**, 347–370. [24]

374. Petes, T.D., and Fangman, W.L. (1973). Preferential synthesis of yeast mitochondrial DNA in α factor-arrested cells. *Biochem. Biophys. Res. Commun.* **55**, 603–609. [64]

375. Pfeiffer, S.E. (1968). RNA synthesis in synchronously growing populations of HeLa S3 cells. II. Rate of synthesis of individual RNA fractions. *J. Cell. Physiol.* **71**, 95–104. [125]

376. Pfeiffer, S.E., and Tolmach, L.J. (1968). RNA synthesis in synchronously growing populations of HeLa S3 cells. I. Rate of total RNA synthesis and its relationship to DNA synthesis. *J. Cell. Physiol.* **71**, 77–94. [124]

377. Phillips, D.M., and Phillips, S.G. (1973). Repopulation of postmitotic nucleoli by preformed RNA. II. Ultrastructure. *J. Cell Biol.* **58**, 54–63. [95]

378. Phillips, S.G. (1972). Repopulation of the postmitotic nucleolus by preformed RNA. *J. Cell Biol.* **53**, 611–623. [95]

379. Pica-Mattoccia, L., and Attardi, G. (1972). Expression of the mitochondrial genome in HeLa cells. IX. Replication of mitochondrial DNA in relationship to the cell cycle in HeLa cells. *J. Mol. Biol.* **64**, 465–484. [63–64]

380. Platz, R.D., Stein, G.S., and Kleinsmith, L.J. (1973). Changes in the phosphorylation of non-histone chromatin proteins during the cell cycle of HeLa S3 cells. *Biochem. Biophys. Res. Commun.* **51**, 735–740. [122]

381. Plaut, W., Nash, D., and Fanning, T. (1966). Ordered replication of DNA in polytene chromosomes of *Drosophila melanogaster. J. Mol. Biol.* **16**, 85–93. [74]

382. Pollack, R.E., and Teebor, G.W. (1969). Relationship of contact inhibition to tumor transplantability, morphology, and growth rate. *Cancer Res.* **29**, 1770–1772. [135]

383. Pollack, R.E., Green, H., and Todaro, G.J. (1968). Growth control in cultured cells: Selection of sublines with increased sensitivity to contact inhibition and decreased tumor-producing ability. *Proc. Natl. Acad. Sci. U.S.A.* **60**, 126–133. [135]

384. Porter, K., Prescott, D., and Frye, J. (1973). Changes in surface morphology of Chinese hamster ovary cells during the cell cycle. *J. Cell Biol.* **57**, 815–836. [25, 97, 99–101]

385. Porter, K.R., Puck, T.T., Hsie, A.W., and Kelley, D. (1974). An electron microscope study of the effects of dibutyryl cyclic AMP on Chinese hamster ovary cells. *Cell* **2**, 145–162. [113]

386. Prescott, D.M., and Murti, K.G. (1974). Chromosome structure in ciliated protozoans. *Cold Spring Harbor Symp. Quant. Biol.* **38**, 609–618. [68]

387. Prescott, D.M., and Kuempel, P.L. (1972). Bidirectional replication of the chromosome in *Escherichia coli. Proc. Natl. Acad. Sci. U.S.A.* **69**, 2842–2845. [73]

388. Prescott, D.M. (1970). The structure and replication of eukaryotic chromosomes. *Adv. Cell Biol.* **1**, 57–117. [95]

389. Prescott, D.M., and Goldstein, L. (1968). Proteins in nucleocytoplasmic interactions. III. Redistributions of nuclear proteins during and following mitosis in *Amoeba proteus. J. Cell Biol.* **39**, 404–414. [94]

390. Prescott, D.M., and Goldstein, L. (1967). Nuclear–cytoplasmic interaction in DNA synthesis. *Science* **155**, 469–470. [64]

391. Prescott, D.M. (1966). The syntheses of total macronuclear protein, histone, and DNA during the cell cycle in *Euplotes eurystomus. J. Cell Biol.* **31**, 1–9. [120]

392. Prescott, D.M. (1964). Cellular sites of RNA synthesis. *Prog. Nucleic Acid Res. Mol. Biol.* **3**, 33–57. [91]

393. Prescott, D.M., and Bender, M.A. (1963). Synthesis and behavior of nuclear proteins during the cell life cycle. *J. Cell. Comp. Physiol.* **62**, 175–194. [119]

394. Prescott, D.M., and Bender, M.A. (1962). Synthesis of RNA and protein during mitosis in mammalian tissue culture cells. *Exp. Cell Res.* **26**, 260–268. [48, 93]

395. Prescott, D.M. (1959). Variations in the individual generation times of *Tetrahymena geleii* HS. *Exp. Cell Res.* **16**, 279–284. [22–23]

396. Prescott, D.M. (1957). Change in the physiological state of a cell population as a function of culture growth and age (*Tetrahymena geleii*). *Exp. Cell Res.* **12**, 126–134. [45]

397. Prescott, D.M. (1956). Relation between cell growth and cell division. II. The effect of cell size on cell growth rate and generation time in *Amoeba proteus. Exp. Cell Res.* **11**, 86–98. [24]

398. Prescott, D.M. (1955). Relations between cell growth and cell division. I. Reduced weight, cell volume, protein content, and nuclear volume of *Amoeba proteus* from division to division. *Exp. Cell Res.* **9**, 328–337. [11, 13–14]

399. Pritchard, R.H., and Lark, K.G. (1964). Induction of replication by thymine starvation at the chromosome origin in *Escherichia coli. J. Mol. Biol.* **9**, 288–307. [78]

400. Rao, M.V.N., and Prescott, D.M. (1970). Inclusion of predivision labeled nuclear RNA in post division nuclei in *Amoeba proteus. Exp. Cell Res.* **62**, 286–292. [95]

401. Rao, P.N., Hittelman, W.N., and Wilson, B.A. (1975). Mammalian cell fusion. VI. Regulation of mitosis in binucleate HeLa cells. *Exp. Cell Res.* **90**, 40–46. [88]

402. Rao, P.N., and Johnson, R.T. (1970). Mammalian cell fusion. I. Studies on the regulation of DNA synthesis and mitosis. *Nature (London)* **225**, 159–164. [64–65, 88]

403. Rasch, E.M., Barr, H.J., and Rasch, R.W. (1971). The DNA content of sperm of *Drosophila melanogaster. Chromosoma* **33**, 1–18. [74, 130]

404. Raska, K., Jr. (1973). Cyclic AMP in G1-arrested BHK21 cells infected with adenovirus type 12. *Biochem. Biophys. Res. Commun.* **50**, 35–41. [116]

405. Rein, A., Carchman, R.A., Johnson, G.S., and Pastan, I. (1973). Simian virus 40 rapidly lowers cAMP levels in mouse cells. *Biochem. Biophys. Res. Commun.* **52**, 899–904. [116]

406. Richards, B.M., and Bajer, A. (1961). Mitosis in endosperm. Changes in nuclear and chromosome mass during mitosis. *Exp. Cell Res.* **22**, 503–508. [94]

407. Rickinson, A.B. (1970). The effects of low concentrations of actinomycin D on the progress of cells through the cell cycle. *Cell Tissue Kinet.* **3**, 335–347. [53–54, 85, 90]

407a. Rieber, M., and Bacalao, J. (1974). Alterations in nuclear phosphoproteins of a temperature-sensitive Chinese hamster cell line exposed to non-permissive conditions. *Exp. Cell Res.* **85**, 334–339. [134]

408. Ringertz, N.R., Carlsson, S-A., Ege, T., and Bolund, L. (1971). Detection of human and chick nuclear antigens in nuclei of chick erythrocytes during reactivation in heterokaryons with HeLa cells. *Proc. Natl. Acad. Sci. U.S.A.* **68**, 3228–3232. [122]

409. Robbins, E., and Scharff, M.D. (1967). The absence of a detectable G$_1$ phase in a cultured strain of Chinese hamster lung cell. *J. Cell Biol.* **34**, 684–688. [38, 51]

410. Robinson, J.H., Smith, J.A., and King, R.J.B. (1974). Regulation of the proliferation rate of cultured, androgen-responsive cells does not involve changes in cyclic AMP levels. *Cell* **3**, 361–365. [112]

411. Roller, B.A., Hirai, K., and Defendi, V. (1974). Effect of cAMP on nucleoside metabolism. II. Cell cycle dependence of thymidine transport. *J. Cell. Physiol.* **84**, 333–342. [113]

412. Ron, A., and Prescott, D.M. (1969). The timing of DNA synthesis in *Amoeba proteus*. *Exp. Cell Res.* **56**, 430–434. [49]

413. Roscoe, D.H., Robinson, H., and Carbonell, A.W. (1973). DNA synthesis and mitosis in a temperature sensitive Chinese hamster cell line. *J. Cell Physiol.* **82**, 333–338. [133–134]

414. Rovera, G., and Baserga, R. (1971). Early changes in the synthesis of acidic nuclear proteins in human diploid fibroblasts stimulated to synthesize DNA by changing the medium. *J. Cell Physiol.* **77**, 201–212. [122]

415. Rozengurt, E., and Jimenez de Asua, L. (1973). Role of cyclic 3':5'-adenosine monophosphate in the early transport changes induced by serum and insulin in quiescent fibroblasts. *Proc. Natl. Acad. Sci. U.S.A.* **70**, 3609–3612. [114]

416. Rubin, H., and Steiner, R. (1975). Reversible alterations in the mitotic cycle of chick embryo cells in various states of growth regulation. *J. Cell. Physiol.* **85**, 261–270. [45]

417. Rubin, R.W., and Everhart, L.P. (1973). The effect of cell-to-cell contact on the surface morphology of Chinese hamster ovary cells. *J. Cell Biol.* **57**, 837–844. [97, 102]

418. Rudkin, G.T. (1965). Nonreplicating DNA in giant chromosomes. *Genetics* **52**, 470. [82]

419. Rudland, P.S., Seeley, M., and Seifert, W. (1974). Cyclic GMP and cyclic AMP levels in normal and transformed fibroblasts. *Nature (London)* **251**, 417–419. [113–115]

420. Rusch, H.P. (1969). Some biochemical events in the growth cycle of *Physarum polycephalum*. *Fed. Proc., Fed. Am. Soc. Exp. Biol.* **28**, 1761–1770. [20]

421. Russell, D.H., and Stambrook, P.J. (1975). Cell cycle specific fluctuations in adenosine 3':5'-cyclic monophosphate and polyamines of Chinese hamster cells. *Proc. Natl. Acad. Sci. U.S.A.* **72**, 1482–1486. [108]

422. Sachs, L. (1974). Regulation of membrane changes, differentiation, and malignancy in carcinogenesis. *Harvey Lect.* **68**, 1–35. [106]

423. Sachsenmaier, W., Remy, U., and Plattner-Schobel, R. (1972). Initiation of synchronous mitosis in *Physarum polycephalum*. *Exp. Cell Res.* **73**, 41–48. [88]

424. Sachsenmaier, W., von Fournier, D., and Gürtler, K.F. (1967). Periodic thymidine kinase production in synchronous plasmodia of *Physarum polycephalum:* Inhibition by actinomycin and actidion. *Biochem. Biophys. Res. Commun.* **27**, 655–660. [55]

425. Sachsenmaier, W., and Ives, D.H. (1965). Periodische Änderungen der Thymidinkinase-aktivität im synchronen Mitosecyclus von *Physarum polycephalum*. *Biochem. Z.* **343**, 399–406. [128]

426. Salas, J., and Green, H. (1971). Proteins binding to DNA and their relation to growth in cultured mammalian cells. *Nature (London), New Biol.* **229**, 165–169. [66]

427. Salb, J.M., and Marcus, P.I. (1965). Translational inhibition in mitotic HeLa cells. *Proc. Natl. Acad. Sci. U.S.A.* **54**, 1353–1358. [93]

428. Sandberg, A.A., Sofuni, T., Takagi, N., and Moore, G.E. (1966). Chronology and pattern of human chromosome replication. IV. Autoradiographic studies of binucleate cells. *Proc. Natl. Acad. Sci. U.S.A.* **56**, 105–110. [62]

429. Scharff, M.D., and Robbins, E. (1966). Polyribosome disaggregation during metaphase. *Science* **151**, 992–995. [93]

430. Scharff, M.D., and Robbins, E. (1965). Synthesis of ribosomal RNA in synchronized HeLa cells. *Nature (London)* **208**, 464–466. [124]

431. Scherbaum, O., and Zeuthen, E. (1954). Induction of synchronous cell division in

mass cultures of *Tetrahymena pyriformis*. *Exp. Cell Res.* **6,** 221–227. [31]

432. Schindler, R., and Schaer, J.C. (1973). Preparation of synchronous cell cultures from early interphase cells obtained by sucrose gradient centrifugation. *Methods Cell Biol.* **6,** 43–65. [21]

433. Schindler, R., Grieder, A., and Maurer, U. (1972). Studies on the division cycle of mammalian cells. VI. DNA polymerase activities in partially synchronous suspension cultures. *Exp. Cell Res.* **71,** 218–224. [57]

434. Schindler, R., Odartchenko, N., Grieder, A., and Ramseier, L. (1968). Studies on the division cycle of mammalian cells. II. Causal relationship between completion of DNA synthesis and onset of the G2 period. *Exp. Cell Res.* **51,** 1–11. [89]

435. Schneiderman, M.H., and Billen, D. (1973). Methylation of rapidly reannealing DNA during the cell cycle of Chinese hamster cells. *Biochim. Biophys. Acta* **308,** 352–360. [82]

436. Schneiderman, M.H., Dewey, W.C., and Highfield, D.P. (1971). Inhibition of DNA synthesis in synchronized Chinese hamster cells treated in G1 with cycloheximide. *Exp. Cell Res.* **67,** 147–155. [52]

437. Schönherr, O.T., and Wanka, F. (1971). An investigation of DNA polymerase in synchronously growing *Chlorella* cells. *Biochim. Biophys. Acta* **232,** 83–93. [57]

438. Schönherr, O.T., Wanka, F., and Kuyper, C.M.A. (1970). Periodic change of deoxyribonuclease activity in synchronous cultures of *Chlorella*. *Biochim. Biophys. Acta* **224,** 74–79. [128]

439. Schumm, D.E., Morris, H.P., and Webb, T.E. (1974). Early biochemical changes in phytohemagglutinin-stimulated peripheral blood lymphocytes from normal and tumor-bearing rats. *Eur. J. Cancer* **10,** 107–113. [118]

440. Scott, R.E., Furcht, L.T., and Kersey, J.H. (1973). Changes in membrane structure associated with cell contact. *Proc. Natl. Acad. Sci. U.S.A.* **73,** 3631–3635. [102–103]

441. Scott, R.E., Carter, R.L., and Kidwell, W.R. (1971). Structural changes in membranes of synchronized cells demonstrated by freeze-cleavage. *Nature (London), New Biol.* **233,** 219–220. [102–103]

442. Seifert, W.E., and Rudland, P.S. (1974). Cyclic nucleotides and growth control in cultured mouse cells: Correlation of changes in intracellular 3′:5′-cGMP concentration with a specific phase of the cell cycle. *Proc. Natl. Acad. Sci. U.S.A.* **71,** 4920–4924. [109–111, 114]

443. Seifert, W.E., and Rudland, P.S. (1974). Possible involvement of cyclic GMP in growth control of cultured mouse cells. *Nature (London)* **248,** 138–140. [110–111, 114]

444. Seki, S., and Mueller, G.C. (1975). A requirement for RNA, protein, and DNA synthesis in the establishment of DNA replicase activity in synchronized HeLa cells. *Biochim. Biophys. Acta* **378,** 354–362. [83–84]

445. Shall, S. (1973). Selection synchronization by velocity sedimentation separation of mouse fibroblast cells grown in suspension culture. *Methods Cell Biol.* **7,** 269–285. [21–22]

446. Shepherd, G.R., Hardin, J.M., and Noland, B.J. (1971). Methylation of lysine residues of histone fractions in synchronized mammalian cells. *Arch. Biochem. Biophys.* **143,** 1–5. [120–121]

447. Sheppard, J.R. (1973). Cyclic AMP and cell division. *In* "Molecular Pathology" (R.A. Good and S. Day, eds.), p. 405–418. Thomas, Springfield, Illinois. [112]

448. Sheppard, J.R. (1972). Difference in the cyclic adenosine 3′,5′-monophosphate levels in normal and transformed cells. *Nature (London), New Biol.* **236,** 14–16. [113]

449. Sheppard, J.R., and Prescott, D.M. (1972). Cyclic AMP levels in synchronized mammalian cells. *Exp. Cell Res.* **75,** 293–296. [107–108]

450. Showacre, J.L., Cooper, W.G., and Prescott, D.M. (1967). Nucleolar and nuclear RNA synthesis during the cell life cycle in monkey and pig kidney cells *in vitro*. *J. Cell Biol.* **33,** 273–279. [124]

451. Simard, A., Corneille, L., Deschamps, Y., and Verly, W.G. (1974). Inhibition of cell proliferation in the livers of hepatectomized rats by a rabbit hepatic chalone. *Proc. Natl. Acad. Sci. U.S.A.* **71**, 1763–1766. [48]

452. Simmons, T., Heywood, P., and Hodge, L.D. (1974). Intranuclear site of replication of adenovirus DNA. *J. Mol. Biol.* **89**, 423–433. [72]

453. Sims, R.T. (1965). The synthesis and migration of nuclear proteins during mitosis and differentiation of cells in rats. *Q. J. Microsc. Sci.* [N.S.] **106**, 229–239. [94]

454. Sisken, J.E., and Iwasaki, T. (1969). The effects of some amino acid analogs on mitosis and the cell cycle. *Exp. Cell Res.* **55**, 161–167. [89]

455. Sisken, J.E., and Kinosita, R. (1961). Timing of DNA synthesis in the mitotic cycle *in vitro. J. Biophys. Biochem. Cytol.* **9**, 509–518. [36]

456. Sisken, J.E., and Wilkes, E. (1967). The time of synthesis and the conservation of mitosis-related proteins in cultured human amnion cells. *J. Cell Biol.* **34**, 97–110. [89]

457. Skoog, K.L., Nordenskjold, B.A., and Bjursell, K.G. (1973). Deoxyribonucleoside-triphosphate pools and DNA synthesis in synchronized hamster cells. *Eur. J. Biochem.* **33**, 428–432. [57]

458. Slor, H., Bustan, H., and Lev, T. (1973). Deoxyribonuclease II activity in relation to cell cycle in synchronized HeLa S3 cells. *Biochem. Biophys. Res. Commun.* **52**, 556–561. [57]

459. Smets, L.A., and DeLey, L. (1974). Cell cycle dependent modulations of the surface membrane of normal and SV40 virus transformed 3T3 cells. *J. Cell. Physiol.* **84**, 343–348. [105–106]

460. Smets, L.A. (1973). Agglutination with Con A dependent on cell cycle. *Nature (London), New Biol.* **245**, 113–115. [106]

461. Smith, B.J., and Wigglesworth, N.M. (1974). Studies on a Chinese hamster line that is temperature-sensitive for the commitment to DNA synthesis. *J. Cell. Physiol.* **84**, 127–134. [133]

462. Smith, B.J., and Wigglesworth, N.M. (1973). A temperature-sensitive function in a Chinese hamster line affecting DNA synthesis. *J. Cell. Physiol.* **82**, 339–348. [133]

463. Smith, B.J., and Wigglesworth, N.M. (1972). A cell line which is temperature-sensitive for cytokinesis. *J. Cell. Physiol.* **80**, 253–260. [134]

464. Smith, B.J. (1970). Light satellite-band DNA in mouse cells infected with polyoma virus. *J. Mol. Biol.* **47**, 101–106. [81]

465. Smith, D., Tauro, P., Schweizer, E., and Halvorson, H. O. (1968). The replication of mitochondrial DNA during the cell cycle in *Saccharomyces lactis. Proc. Natl. Acad. Sci. U.S.A.* **60**, 936–942. [64]

466. Smith, H.H. (1972). Plant genetic tumors. *Prog. Exp. Tumor Res.* **15**, 138–164. [137]

467. Socher, S.H., and Davidson, D. (1970). Heterogeneity in G_2 duration during lateral root development. *Chromosoma* **31**, 478–484. [88]

468. Solter, D., Škreb, N., and Damjanov, I. (1971). Cell cycle analysis in the mouse egg-cylinder. *Exp. Cell Res.* **64**, 331–334. [51]

469. Sören, L. (1970). Nuclear and cytoplasmic protein accumulation in PHA-stimulated human lymphocytes during blastogenesis. *Exp. Cell Res.* **59**, 244–248. [122]

470. Spardi, S., and Weissbach, A. (1974). The interrelation between DNA synthesis and various DNA polymerase activities in synchronized HeLa cells. *J. Mol. Biol.* **86**, 11–20. [57]

471. Spear, B.B., and Gall, J.G. (1973). Independent control of ribosomal gene replication in polytene chromosomes of *Drosophila melanogaster. Proc. Natl. Acad. Sci. U.S.A.* **70**, 1359–1363. [82]

472. Stambrook, P.J. (1974). The temporal replication of ribosomal genes in synchronized Chinese hamster cells. *J. Mol. Biol.* **82**, 303–313. [82]

473. Stambrook, P.J., and Sisken, J.E. (1972). Induced changes in the rates of uridine-^{3}H uptake and incorporation during the G_1 and S periods of synchronized Chinese

hamster cells. *J. Cell Biol.* **52**, 514–525. [125]

474. Stein, G., and Baserga, R. (1972). Nuclear proteins and the cell cycle. *Adv. Cancer Res.* **15**, 287–330. [122]

475. Stein, G., and Baserga, R. (1970). Continued synthesis of non-histone chromosomal proteins during mitosis. *Biochem. Biophys. Res. Commun.* **41**, 715–722. [93, 122]

476. Stein, G.S., and Matthews, D.E. (1973). Nonhistone chromosomal protein synthesis: Utilization of preexisting and newly transcribed messenger RNA's. *Science* **181**, 71–73. [93]

477. Stein, S.M., and Berestecky, J.M. (1974). Exposure of an arginine-rich protein at surface of cells in S, G_2, and M phases of the cell cycle. *J. Cell Physiol.* **85**, 243–250. [103]

478. Stellwagen, R.H., and Cole, R.D. (1969). Histone biosynthesis in the mammary gland during development and lactation. *J. Biol. Chem.* **244**, 4878–4887. [120]

479. Stenman, S., Zeuthen, J., and Ringertz, N.R. (1975). Expression of SV40 T antigen during the cell cycle of SV40-transformed cells. *Int. J. Cancer* **15**, 547–554. [95]

480. Stevens, A.R., and Prescott, D.M. (1971). Reformation of nucleolus-like bodies in the absence of post-mitotic RNA synthesis. *J. Cell Biol.* **48**, 443–454. [95]

481. Steward, D.L., Shaeffer, J.R., and Humphrey, R.M. (1968). Breakdown and assembly of polyribosomes in synchronized Chinese hamster cells. *Science* **161**, 791–793. [93]

482. Stubblefield, E., and Murphree, S. (1967). Synchronized mammalian cell cultures. II. Thymidine kinase activity in colcemid synchronized fibroblasts. *Exp. Cell Res.* **48**, 652–656. [55–56]

483. Stubblefield, E., Klevecz, R., and Deaven, L. (1967). Synchronized mammalian cell cultures. I. Cell replication cycle and macromolecular synthesis following brief colcemid arrest of mitosis. *J. Cell. Physiol.* **69**, 345–354. [124]

484. Studzinski, G.P., and Lambert, W.C. (1969). Thymidine as a synchronizing agent. I. Nucleic acid and protein formation in synchronous HeLa cultures treated with excess thymidine. *J. Cell. Physiol.* **73**, 109–118. [32]

485. Tapiero, H., Shaool, D., Monier, M.N., and Harel, J. (1974). Replication of repetitious DNA in synchronized chick fibroblast cells. *Exp. Cell Res.* **89**, 39–46. [82]

486. Taylor, E.W. (1963). Relation of protein synthesis to the division cycle in mammalian cell cultures. *J. Cell Biol.* **19**, 1–18. [89]

487. Taylor, J.H., Myers, T.L., and Cunningham, H.L. (1971). Programmed synthesis of deoxyribonucleic acid during the cell cycle. *In Vitro* **6**, 309–321. [79]

488. Taylor, J.H. (1968). Rates of chain growth and units of replication in DNA of mammalian chromosomes. *J. Mol. Biol.* **31**, 579–594. [75]

489. Taylor, J.H. (1960). Asynchronous duplication of chromosomes in cultured cells of Chinese hamster. *J. Biophys. Biochem. Cytol.* **7**, 455–464. [74]

490. Taylor-Papadimitriou, J. (1974). Effects of adenosine 3'5'-cyclic monophosphoric acid on the morphology, growth and cell cycle of Earle's L cells. *Int. J. Cancer* **13**, 404–411. [112]

491. Teel, R.W., and Hall, R.G. (1973). Effect of dibutyryl cyclic AMP on the restoration of contact inhibition in tumor cells and its relationship to cell density and the cell cycle. *Exp. Cell Res.* **76**, 390–394. [113]

492. Temin, H.M. (1971). Stimulation by serum of multiplication of stationary chicken cells. *J. Cell. Physiol.* **78**, 161–170. [45]

493. Terasima, T., Fujiwara, Y., Tanaka, S., and Yasukawa, M. (1968). Synchronous culture of L cells and initiation of DNA synthesis. *In* "Cancer Cells in Culture" (H. Katsuta, ed.), pp. 73–84. Univ. Park Press, Baltimore, Maryland. [52]

494. Terasima, T., and Yasukawa, M. (1966). Synthesis of G1 protein preceding DNA synthesis in cultured mammalian cells. *Exp. Cell Res.* **44**, 669–672. [52–53]

495. Thilly, W.G., Arkin, D.I., Nowak, T.S., Jr., and Wogan, G.N. (1975). Behavior of

subcellular marker enzymes during the HeLa cell cycle. *Biotechnol. Bioeng.* **17,** 695–701. [56, 128]

496. Thompson, L.H., and Baker, R.M. (1973). Isolation of mutants of cultured mammalian cells. *Methods Cell Biol.* **6,** 209–282. [133]

497. Thompson, L.R., and McCarthy, B.J. (1968). Stimulation of nuclear DNA and RNA synthesis by cytoplasmic extracts *in vitro. Biochem. Biophys. Res. Commun.* **30,** 166–172. [66]

498. Tobey, R.A. (1973). Production and characterization of mammalian cells reversibly arrested in G_1 by growth in isoleucine-deficient medium. *Methods Cell Biol.* **6,** 67–112. [33, 54]

499. Tobey, R.A., and Crissman, H.A. (1972). Use of flow microfluorometry in detailed analysis of effects of chemical agents on cell cycle progression. *Cancer Res.* **32,** 2726–2732. [39]

500. Tobey, R.A., Petersen, D.F., and Anderson, E.C. (1971). Biochemistry of G_2 and mitosis. *In* "The Cell Cycle and Cancer" (R. Baserga, ed.), pp. 309–353. Dekker, New York. [90]

501. Tobey, R.A., Anderson, E.C., and Petersen, D.F. (1967). The effect of thymidine on the duration of G_1 in Chinese hamster cells. *J. Cell Biol.* **35,** 53–59. [43]

502. Tobey, R.A., Petersen, D.F., Anderson, E.C., and Puck, T.T. (1966). Life cycle analysis of mammalian cells. III. The inhibition of division in Chinese hamster cells by puromycin and actinomycin. *Biophys. J.* **6,** 567–581. [89]

503. Tobia, A.M., Schildkraut, C.L., and Maio, J.J. (1971). DNA replication in synchronized cultured mammalian cells. III. Relative times of synthesis of mouse satellite and main band DNA. *Biochim. Biophys. Acta* **246,** 258–262. [81]

504. Tobia, A.M., Schildkraut, C.L., and Maio, J.J. (1970). Deoxyribonucleic acid replication in synchronized cultured mammalian cells. I. Time of synthesis of molecules of different average guanine + cytosine content. *J. Mol. Biol.* **54,** 499–515. [80–81]

505. Todaro, G.J., Lazar, G.K., and Green, H. (1966). The initiation of cell division in a contact-inhibited mammalian cell line. *J. Cell. Comp. Physiol.* **66,** 325–334. [44]

506. Todaro, G.J., and Green, H. (1963). Quantitative studies of the growth of mouse embryo cells in culture and their development into established lines. *J. Cell Biol.* **17,** 229–313. [44]

507. Todo, A., Strife, A., Fried, J., and Clarkson, B.D. (1971). Proliferative kinetics of human hematopoietic cells during different growth phases *in vitro.* Cancer Res. **31,** 1330–1340. [44]

508. Tolnai, S. (1965). An analysis of the life cycle of Ehrlich ascites tumor cells. *Lab. Invest.* **14,** 701–710. [44]

509. Tremblay, G.Y., Daniels, M.J., and Schaechter, M. (1969). Isolation of a cell membrane-DNA-nascent RNA complex from bacteria. *J. Mol. Biol.* **40,** 65–76. [68]

510. Troy, M.R., and Wimber, D.E. (1968). Evidence for a constancy of the DNA synthetic period between diploid–polyploid groups in plants. *Exp. Cell Res.* **53,** 145–154. [85]

511. Tsuboi, A., and Baserga, R. (1972). Synthesis of nuclear acidic proteins in density-inhibited fibroblasts stimulated to proliferate. *J. Cell. Physiol.* **80,** 107–118. [35, 44]

512. Utakoji, T., and Hsu, T.C. (1965). DNA replication patterns in somatic and germ-line cells of the male Chinese hamster. *Cytogenetics* **4,** 295–315. [81]

513. van den Biggelaar, J.A.M. (1971). Timing of the phases of the cell cycle during the period of asynchronous division up to the 49-cell stage in *Lymnaea. J. Embryol. Exp. Morphol.* **26,** 367–391. [47, 51]

514. Van't Hof, J. (1970). Metabolism and the prolonged retention of cells in the G1 and S period of the mitotic cycle of cultured pea roots. *Exp. Cell Res.* **61,** 173–182. [43]

515. Van't Hof, J., Hoppin, D.P., and Yagi, S. (1973). Cell arrest in G1 and G2 of the mitotic cycle of *Vicia faba* root meristems. *Am. J. Bot.* **60,** 889–895. [88]

516. Van Wijk, R., Wicks, W.D., Bevers, M.M., and Van Rijn, J. (1973). Rapid arrest of DNA synthesis by $N^6,O^{2'}$-dibutyryl cyclic adenosine $3',5'$-monophosphate in cultured hepatoma cells. *Cancer Res.* **33**, 1331–1338. [113]

517. Verly, W.G., Deschamps, Y., Pushpathadam, J., and Desrosiers, M. (1971). The hepatic chalone. I. Assay method for the hormone and purification of the rabbit liver chalone. *Can. J. Biochem.* **49**, 1376. [48]

518. Volpe, P., and Eremenko, T. (1973). Nuclear and cytoplasmic DNA synthesis during the mitotic cycle of HeLa cells. *Eur. J. Biochem.* **32**, 227–232. [63]

519. Walters, R.A., Tobey, R.A., and Ratliff, R.L. (1973). Cell-cycle-dependent variations of deoxyribonucleoside triphosphate pools in Chinese hamster cells. *Biochem. Biophys. Acta* **319**, 336–347. [57]

520. Walther, W.G., and Edmunds, L.N. (1970). Periodic increase in deoxyribonuclease activity during the cell cycle in synchronized *Euglena. J. Cell Biol.* **46**, 613–617. [128]

521. Wang, R.J. (1974). Temperature-sensitive mammalian cell line blocked in mitosis. *Nature (London)* **248**, 76–78. [134]

522. Wanka, F., and Moors, J. (1970). Selective inhibition by cycloheximide of nuclear DNA synthesis in synchronous cultures of *Chlorella. Biochem. Biophys. Res. Commun.* **41**, 85–90. [83]

523. Webster, P.L., and Van't Hof, J. (1970). DNA synthesis and mitosis in meristems: Requirements for RNA and protein synthesis. *Am. J. Bot.* **57**, 130–139. [88]

524. Weinberg, R., and Penman, S. (1969). Metabolism of small molecular weight monodisperse nuclear RNA. *Biochim. Biophys. Acta* **190**, 10–29. [125]

525. Weinstein, Y., Chambers, D.A., Bourne, H.R., and Melmon, K.L. (1974). Cyclic GMP stimulates lymphocyte nucleic acid synthesis. *Nature (London)* **251**, 352–353. [115]

526. Weintraub, H. (1972). A possible role for histone in the synthesis of DNA. *Nature (London)* **240**, 449–453. [75, 83–84]

527. Weintraub, H., and Holtzer, H. (1972). Fine control of DNA synthesis in developing chick red blood cells. *J. Mol. Biol.* **66**, 13–35. [83]

528. Weiss, B., Shein, H.M., and Snyder, R. (1971). Adenylate cyclase and phosphodiesterase activity of normal and SV40 virus-transformed hamster astrocytes in cell culture. *Life Sci.* **10**, 1253–1260. [116]

529. Weiss, M.C., Todaro, G.J., and Green, H. (1968). Properties of a hybrid between lines sensitive and insensitive to contact inhibition of cell division. *J. Cell. Physiol.* **71**, 105–108. [137]

530. Werry, P.A.T.J., and Wanka, F. (1972). The effect of cycloheximide on the synthesis of major and satellite DNA components in *Physarum polycephalum. Biochim. Biophys. Acta* **287**, 232–235. [82]

531. Wiener, F., Klein, G., and Harris, H. (1974). The analysis of malignancy by cell fusion. VI. Hybrids between different tumor cells. *J. Cell Sci.* **16**, 189–198. [137]

532. Wiener, F., Klein, G., and Harris, H. (1974). The analysis of malignancy by cell fusion. V. Further evidence of the ability of normal diploid cells to suppress malignancy. *J. Cell Sci.* **15**, 177–183. [137]

533. Wiener, F., Klein, G., and Harris, H. (1971). The analysis of malignancy by cell fusion. III. Hybrids between diploid fibroblasts and other tumor cells. *J. Cell Sci.* **8**, 681–692. [137]

534. Wilkinson, L.E., and Pringle, J.R. (1974). Transient G_1 arrest of *S. cerevisiae* cells of mating type α by a factor produced by cells of mating type *a. Exp. Cell Res.* **89**, 175–187. [5]

535. Williams, C.A., and Ockey, C.H. (1970). Distribution of DNA replicator sites in mammalian nuclei after different methods of cell synchronization. *Exp. Cell Res.* **63**, 365–372. [67]

536. Williamson, D.H. (1973). Replication of the nuclear genome in yeast does not require concomitant protein synthesis. *Biochem. Biophys. Res. Commun.* **52**, 731–740. [53, 84]

537. Williamson, D.H., and Moustacchi, E. (1971). The synthesis of mitochondrial DNA during the cell cycle in the yeast *Saccharomyces cerevisiae*. *Biochem. Biophys. Res. Commun.* **42**, 195–201. [63]

538. Willingham, M.C., and Pastan, I. (1975). Cyclic AMP modulates microvillus formation and agglutinability in transformed and normal mouse fibroblasts. *Proc. Natl. Acad. Sci. U.S.A.* **72**, 1263–1267. [117]

539. Willingham, M.C., and Pastan, I. (1974). Cyclic AMP mediates the Concanavalin A agglutinability of mouse fibroblasts. *J. Cell Biol.* **63**, 288–294. [117]

540. Willingham, M.C., Johnson, G.S., and Pastan, I. (1972). Control of DNA synthesis and mitosis in 3T3 cells by cyclic AMP. *Biochem. Biophys. Res. Commun.* **48**, 743–748. [88, 116]

541. Wise, G.E., and Prescott, D.M. (1973). Initiation and continuation of DNA replication are not associated with the nuclear envelope in mammalian cells. *Proc. Natl. Acad. Sci. U.S.A.* **70**, 714–717. [68–69]

542. Wolstenholme, D.R. (1973). Replicating DNA molecules from eggs of *Drosophila melanogaster*. *Chromosoma* **43**, 1–18. [78]

543. Xeros, N. (1962). Deoxyriboside control and synchronization of mitosis. *Nature (London)* **194**, 682–683. [32]

544. Yamaguchi, N., and Weinstein, I.B. (1975). Temperature-sensitive mutants of chemically transformed epithelial cells. *Proc. Natl. Acad. Sci. U.S.A.* **72**, 214–218. [139]

545. Yamamura, H., Kumon, A., Nishiyama, K., Takeda, M., and Nishizuka, Y. (1971). Characterization of two adenosine 3′,5′-monophosphate-dependent protein kinases from rat liver. *Biochem. Biophys. Res. Commun.* **45**, 1560–1566. [118]

546. Yang, D.-P., and Dodson, E.O. (1970). The amounts of nuclear DNA and the duration of DNA synthetic period(s) in related diploid and autotetraploid species of oats. *Chromosoma* **31**, 309–320. [85]

547. Yasumasu, I., Fujiwara, A., and Ishida, K. (1973). Periodic change in the content of adenosine 3′5′-cyclic monophosphate with close relation to the cycle of cleavage in the sea urchin egg. *Biochem. Biophys. Res. Commun.* **54**, 628–632. [108–109]

548. Yoshikura, H., and Hirokawa, Y. (1968). Induction of cell replication. *Exp. Cell Res.* **52**, 439–444. [44]

549. Young, R.W. (1962). Regional differences in cell generation time in growing rat tibiae. *Exp. Cell Res.* **26**, 562–567. [46]

550. Zeilig, C.E., Johnson, R.A., Sutherland, E.W., and Friedman, D.L. (1974). Cyclic AMP levels in synchronized HeLa cells and a dual effect on mitosis. *Fed. Proc., Fed. Am. Soc. Exp. Biol.* **33**, 1391. [108]

551. Zellweger, A., Ryser, U., and Braun, R. (1972). Ribosomal genes of *Physarum:* Their isolation and replication in the mitotic cycle. *J. Mol. Biol.* **64**, 681–691. [82]

552. Zetterberg, A. (1966). Synthesis and accumulation of nuclear and cytoplasmic proteins during interphase in mouse fibroblasts *in vitro*. *Exp. Cell Res.* **42**, 500–511. [122]

553. Zetterberg, A., and Killander, D. (1965). Quantitative cytochemical studies on interphase growth. II. Derivation of synthesis curves from the distribution of DNA, RNA and mass values of individual mouse fibroblasts *in vitro*. *Exp. Cell Res.* **39**, 22–32. [124]

554. Zetterberg, A., and Killander, D. (1965). Quantitative cytophotometric and autoradiographic studies on the rate of protein synthesis during interphase in mouse fibroblasts *in vitro*. *Exp. Cell Res.* **40**, 1–11. [16–17]

555. Zeuthen, E. (1971). Synchronization of the *Tetrahymena* cell cycle. *Adv. Cell Biol.* **2**, 111–152. [31]

556. Zeuthen, E., ed. (1964). "Synchrony in Cell Division and Growth." Wiley (Interscience), New York. [34]

557. Zeuthen, E. (1961). The cartesian diver balance. *Gen. Cytochem. Methods* **2**, 61–91. [10]

558. Zeuthen, E. (1953). Growth as related to the cell cycle in single-cell cultures of *Tetrahymena pyriformis*. *J. Embryol. Exp. Morphol.* **1**, 239–249. [14–15]

559. Zeuthen, E. (1948). A cartesian diver balance weighing reduced weights (R.W.) with an accuracy of ± 0.01 micrograms. *C. R. Trav. Lab. Carlsberg* **26**, 243–266. [10]

560. Zylber, E.A., and Penman, S. (1971). Synthesis of 5S and 4S RNA in metaphase-arrested HeLa cells. *Science* **172**, 947–949. [51, 91–92]

Index

A

Acid phosphatases, 128
Actinomycin D
  cell cycle inhibition by, 49
  effect on thymidine kinase synthesis, 56
Adenovirus DNA, replication of, 72
Adenovirus type, 12, 116
Adenyl cyclase, 109–110, 114, 116–117, 137
ADP, reduction of, 57
Agglutinability
  of cells during the cell cycle, 105–106
  of transformed cells, 105, 117
Alkaline phosphatases, 128
Amethopterin, 52
  arrest of cells in S with, 89
  inhibition of TMP synthesis, 102
Amino acid deprivation, 33–34, 54
Amnion cells, 67
Amoeba
  arrest in G_2, 50
  cell cycle of, 11, 13
  cell size, 24
  DNA synthesis in, 50
  generation time in, 24
  growth curves for, 13–14
  multinucleated, 60
  nuclear volume of, 13
  unequal division of, 11
Amoeba proteus, 22, 49
  cell cycle of, 49–50

  growth in mass of, 10
cAMP, *see* Cyclic AMP
Amphiuma means, chromosomes of, 73
Ascites tumor cells, 47, 57, 121
  binucleated, 62
  cell cycle of, 43
  lack of G_1 in, 51
  multinucleated, 62
Aspartate transcarbamoylase, 128
dATP, pool of, 32
Avena, S period in, 85

B

Bean roots, S period in, 85
BHK21 cells, arrested in G_1, 116
Binucleated mammalian cells, 61
Blebs, 97

C

Caffeine, 109
  multinucleated cells induced by, 61
Canavanine, 84
Carbamoylphosphate synthetase, 128
Carbohydrate starvation, 43
Carcinogen(s), 135, 137, 138
Cartesian diver balance, 10–12, 24
Cartesian diver respirometer, 14–15
Cartilage, 47
Catalase, 128

Cell cycle
    cAMP and, 107
    genetic analysis in mammalian cells, 133
    genetics of, 130
    major features of, 3
    sections of, 2
Cell cycle analysis, 34
Cell cycle genes, 131
Cell division
    activities during, 9
    time taken for, 29
    unequal, 28
Cell fusion, 34, 137
Cell growth
    during the cell cycle, 7
    relation to cell division, 7
Cell mass, variableness in, 39
Cell reproduction
    components of, 2
    control of in G_1, 44
    nonspecific regulation, 4
    regulation of, 3, 107
    regulatory genes for, 137
    specific regulation, 4
Cell selection, individual, 22
Cell size, of daughter cells, 28
Cell surface changes, 97
Cell volume(s), distribution of, 28
Chalone(s), 4, 47, 117–118
    G_1, 48
    G_2, 48
    liver cell, 48
    lymphocyte, 48
    rat liver, 48
Chick cells, 117
    release from stationary phase, 45
Chicken embryo cells, 103
Chick erythrocyte nuclei
    reactivation of DNA synthesis in, 122
    reactivation of RNA synthesis in, 122
Chick erythrocytes, fused with HeLa, 61
Chick fibroblasts
    embryo, 114, 116
    mitochondrial DNA synthesis in, 63
Chinese hamster cells, scanning electron
    micrograph of, 25
Chlamydomonas reinhardtii, ts mutants of,
    133
Chlorella, 83, 128
Chromatin
    reconstituted, 91

    RNA synthesis in, 45
    template activity, 45
Chromatin proteins, phosphorylation of
    nonhistone, 122–123
Chromatin structure, 120
Chromosomal proteins, rate of synthesis of,
    93
Chromosome(s)
    multiple replicons in, 74
    premature condensation of, 87
    proteins of mitotic, 91
Chromosome condensation, 87, 120–121
Chromosome condensation factor(s), 88
Ciona, 130–131
Colcemid, thymidine kinase during block
    with, 56
Colon, 46
Concanavalin A, 105–106
    binding in mouse 3T3 cells, 106
Crypt cells, 47
dCTP, formation of, 32
Cyclic AMP, 46, 107
    and G_2 arrest, 88
    changes in, 111
    changes in intracellular concentration, 108
    increase in, 113
    increase in G_1, 107
    induced increases in, 112
    serum-reduced reduction in, 114
Cyclic AMP-dependent protein kinase, 113
Cyclic AMP levels, regulation of, 117
Cyclic GMP, 107
    changes in, 111
    concentration of, 109
    ratio to cyclic AMP, 115
    rise in, 110
Cycloheximide, 53, 56, 83
Cytidine nucleotide, reduction of, 57
Cytochalasin B, 61
Cytochalasin-produced binucleates, 61
Cytokinesis
    inhibition of in amoeba, 11
    ts mutant for, 134
Cytoplasmic factors, in DNA synthesis, 67
Cytoplasm–nucleus ratio, 7
Cytosine arabinoside, 31–32, 33

D

D period, definition of, 3
Daughter cells, mass of individual, 39

Density-dependent cells, 45
  infection of inhibited, 116
  inhibited, 115
Density-dependent inhibition, 37, 44–45,
    103, 111, 114, 117, 122, 135–138
  release by insulin, 114
  release from, 35
  release of animal cells from, 34
3'–Deoxyadenosine inhibition of RNA
    synthesis, 90
Deoxycytidine deaminase, 57
Deoxycytidine kinase, 57
Deoxycytidylate, 57
Deoxynucleoside triphosphate(s), synthesis
    of, 55
Deoxynucleoside triphosphate pools, sizes
    of, 57
Dibutyryl-cAMP
  effect on growth, 112
  inhibition of thymidine uptake by, 113
Differentiated cells, amount of DNA in, 46
Dipodymus ordii, 81
DNA
  replication of highly repetitious, 81
  replication of the viral, 116
DNA-binding proteins, 124
DNA fiber autoradiography, 74
DNA membrane complexes, 68
DNA polymerase
  activity during the cell cycle, 57
  and initiation of the S period, 57
  inhibition of, 31–32
  relation between DNA synthesis and, 58
  shift between nucleus and cytoplasm, 20
  shift from cytoplasm to nucleus, 58
  of *Xenopus,* 67
DNA replication machinery, 68, 72
DNA synthesis
  in amoeba, 49–50
  in *Amoeba proteus,* 49
  asynchrony in mitochondria, 63
  bidirectional, 75
  block of, 32
  in cell cycle recombinants, 64
  in cell-free systems, 66
  and cell mass, 40–41
  cell size and, 42
  continuation of, 68
  decline following the inhibition of protein
      synthesis, 83
  deprivation of other amino acids and, 84

differential, 82
  effect of cycloheximide on, 83
  in eukaryotic cells, 74
  events leading to, 33–34
  in hamster cells, 27
  histone synthesis and, 84
  histone synthesis during, 119
  inducer of, 66
  induction in G_1 nuclei, 65
  inhibition of, 31–32, 33, 83
  initiating signal(s) for, 65
  initiation by cytoplasmic extracts, 67
  initiation of, 41–43, 46, 49, 68, 72, 115,
      133
  intranuclear site of initiation of, 67
  and isolated nuclear membranes, 68
  in isolated nuclei, 66
  macronuclear, 63
  micronuclear, 63
  mitochondrial, 64
  mitosis and, 49
  nuclear-cytoplasmic interactions in, 60
  ordering of, 78, 80
  origin of, 68
  preparation for, 55
  protein synthesis and, 42, 52
  in prokaryotic cells, 73
  protein synthesis necessary for initiation
      of, 53–54
  proteins needed to initiate, 50
  relationship of protein synthesis to, 83
  ribosomal RNA synthesis and, 85
  RNA and, 52
  RNA synthesis and the continuation of,
      84
  RNA synthesis necessary for the initiation
      of, 53
  in slime mold, 79
  synchronized in mitochondria, 64
  synchrony in the initiation of, 61
  in *Triturus,* 76
  tryptophan deprivation and, 84
DNA viruses, oncogenic, 116
rDNA, replication of, 82
DNase, 128
  alkaline, 128
Double thymidine block, 32
Drosophila, 130–131, 136, 138
  chromosomes in, 74
  DNA replication in, 74
  rate of DNA synthesis in, 78

replicating DNA from, 77
S period in, 78
Drosophila melanogaster
  DNA in largest chromosomes in, 73
Duodenum, 46

E

Ear epidermis, 88
Ecdysone, 129
Embryogenesis, cell reproduction in, 47
Endoreduplication, 52, 131
Endosperm cells, 94
Enzyme activities, patterns of, 127
Epidermal tissue, cell proliferation in, 117
Epinephrine, 117
Epithelial cells
  generation time of, 46
  mouse, 47
  proliferation of, 46
Erythrocyte(s)
  average life time of, 1
  number in humans, 1
  plasma membrane of, 117
  rate of production of, 1
Escherichia coli, 40
  circular DNA molecule in, 73
  replication of the chromosome in, 79
  RNA polymerase from, 45
Esophagus, 46
Euchromatin, synthesis of, 81
Euglena, 128
Euplotes, 22, 61–62
  histone synthesis in, 120
  macronuclear G_1, 49
  macronucleus
    autoradiography of, 71
  micronucleus of, 49
Euplotes eurystomus
  macronucleus of, 68, 70
  replication bands of, 70

F

F_1 histone, 120
  phosphokinase, 121
  phosphorylation of, 121
Fibroblast
  cell cycle of, 16
  dry mass increase during the cell cycle, 16
  growth of, 16
  rate of protein synthesis, 17

Ficoll density gradient, 40–41
Filapodia, in G_2, 97
Fish embryos, 95
Fish hybrid(s), 138
Fluorodeoxyuridine, 31–32, 52
  release from inhibition by, 31
Freeze-fracture, 103
Frog cell, haploid, *see* Haploid frog cell
FUdR, 33
Fumarase, 127–128
Fused cells, DNA synthesis in, 61

G

G_0 arrest, by serum deprivation, 41
G_0 state, 3, 6, 36, 38, 45, 67, 110
  entry into, 48
  reversal of, 34, 115
  ribosomal RNA synthesis for reversal of, 53
G_1 arrest, 44, 118
  by serum deprivation, 41
  cyclic nucleotides and, 113
  release from, 114–116
G_1 events, 51
G_1-less cells, in multicellular organisms, 51
G_1-less hamster cell line, 38, 51
G_1 period
  absence of, 49
  amino acid deprivation, sensitivity of, 54
  cAMP, rise of, 107
  arrest in, 3, 34, 37, 45
    prolonged, of tissue cells, 48
    reversible, of cells in, 44
    reversible, of tissue cells in, 46
  average, determination of, 30
  average lengths of, 47
  blocking conditions, 46
  cell growth to length of, relation of, 39
  checkpoint in, 37–38, 46
  cultured cells lacking, 51
  definition of, 3
  extension of, 43
  flexibility in, 43
  length of, 40
    variability of, 28, 30, 36–39, 43
  mass increase during, 39–40
  progress, erasure of, 53
  protein synthesis in, 53
  requirements of, 52
  RNA synthesis, requirements for, 52

synchronization in, 33
G$_2$ period
  absence of, 87
  cAMP in, 109
  cAMP-sensitive point in, 116
  arrest in, 49, 88
  blockage of transition from, 89
  definition of, 3
  determination of average, 30
  length of, 29
  proteins synthesized in, 90
    requirements for, 89
  ribosomal RNA and, 89
  RNA syntheses, requirements for, 89
  thymidine kinase activity during, 56
  variation in, 30
GDP, reduction of, 57
Generation time(s)
  definition of, 2
  determination of average, 30
  distributions in, 24, 27
  estimate of, 29
  length of, variableness, 37
  variableness in, 36
  variation in, 27–28
Glucocorticoid hormones, 128
Glucose-6-phosphate dehydrogenase,
  127–128
cGMP, *see* Cyclic GMP
Grasshopper, neuroblasts in, 51
dGTP, pool of, 32

H

Hamster cell(s)
  electron microscope autoradiograph of, 69
  mitochondrial DNA synthesis in, 63
Hamster cheek pouch, epithelium of, 47
Hamster fibroblasts, G$_1$-less cycle in, 51
Hamster K-12 cells, temperature-sensitive
  mutant of, 133
Hamster ovary cells, cAMP in, 108
Haploid frog cells, cell cycle of, 41
Heat shocks, for synchrony, 31
HeLa cell(s), 61
  mRNA in, 93
  mitochondrial DNA synthesis in, 63
HeLa cell fusions, 65
  G$_2$ delay in, 88
Heparan sulfate, 106
Hepatocytes, 48

Hepatoma cells, 112, 128
Heterochromatic chromosomes, nonreplica-
  tion of, 82
Heterochromatin
  attachment of, 72
  DNA replication, exclusion of, 82
  synthesis of, 67, 81
Heterochromatin DNA, replication of, 72
Heterokaryons, 61
  HeLa–hamster, 61
  HeLa–mouse, 61
Histone(s)
  acetylations of, 121
  doubling in, 119
  genes for, 119
  mRNA for, 119
    destruction of, 119
  phosphorylation of F$_1$, 90
  transcription of, 119
Histone acetylation, 120
Histone methylation, 120
Histone modification, 120
Histone phosphorylation, 120
Histone synthesis, 84, 119
  in the absence of DNA synthesis, 119
  inhibition by cycloheximide, 83
HnRNA, 94, 126
Human fibroblasts, 45
  diploid, 113
Humans, cell number in, 1
Hybrid(s)
  human–mouse cell, 80
  mouse–hamster, 62, 79
Hybrid cells, replication patterns of chromo-
  somes in, 79
Hydra
  cell cycle in, 51
  epithelium of, 51
Hydroxyurea, 31–32, 34, 41
Hydroxyurea blockpoint, 134

I

Ileum, 46
Interferon, 116
Intramembranous particles, 103
  aggregation of, 103
  population density of, 102
Isoleucine deprivation, 33–34, 54
  mitochondrial DNA synthesis and, 63
Isoproterenol, 122

K

Kangaroo rat, 81
  cells of, 76
Kidney, 48

L

Labeled mitotic index method, 29–30
Lactic dehydrogenase, 127–128
Lactuca sativa, 49
Lectins, 104–106, 117, 136
Leucine, deprivation of, 33, 54
Leukemic lymphocytes, 48
Liver, 48
  inhibitor from, 44
  mitochondrial DNA synthesis in regenerating, 63
Liver cells
  human, 112
  mitochondrial DNA synthesis in, 63
Lymphocytes, 43, 46
  cGMP in, 114
  human, 47
  number of, 2
  of peripheral blood, 48
  production rate of, 2
Lymphoid cell line, 107
  cAMP in, 109
Lymphoma cells, 43
  mitochondrial DNA synthesis in, 63
Lymphoma lymphocytes, 48

M

M-band, 72
  technique, 68
Macromelanophore(s), 136
  genes for regulating reproduction of, 137
Macronuclei
  synchronous DNA synthesis in, 60
Macronucleus, 42
  DNA synthesis in, 62
Mammalian chromosome, DNA of, 73
Mammalian S period, duration of, 76
Mammary carcinoma cultures, 112
Mating factor, in yeast, 133
Mealy bug, 82
Melanoma(s), 137–138
Membrane proteins, change in, 104
Mesenchyme cells, 46
Methotrexate, 31

Micronucleus, DNA synthesis in, 62
Microvilli, 103
  during G_2, 97
Mitosis
  cAMP, decrease in, 107
  preparation for, 87
  protein synthesis during, 93
  proteins, migration of, 94
  RNA synthesis during, 91
  4 S RNA, synthesis of, 91
  5 S RNA, synthesis of, 91
  thymidine kinase activity during, 56
Mitotic cells
  synchrony of, 27
  variable size of, 39
  volume distribution of, 28
Mitotic selection, 24, 29–30, 33, 37–39
  entry into DNA synthesis after, 38
  of hamster cells, 26
Monkey cells, 113
Mouse embryos, 51
Mouse LS cells, synchronous culture of, 22
Mouse lymphoblasts, 115
Mouse lymphoma cells, 113
Mutagens, 135
Myeloid-erythroid series, lack of G_1 in, 51

N

Neuroblastoma, 138
  genetic basis of, 136
Neuroblasts, 51, 136
Neurons, 48
Nicotiana, hybrid crosses of, 137
NP-40, 126
NRK cells, 114
Nuclear envelope, 68–69, 71
  and DNA synthesis, 67
  fragmentation of, 94
  purified from mouse cells, 72
Nuclear membranes, analysis of purified, 72
Nuclear proteins, 119
  nonhistone, 122
  phosphorylation of, 118
  release to the cytoplasm of, 95
Nuclear RNA, release during mitosis, 94–95
Nuclear transplantation, 34
Nucleolus
  breakdown of, 94
  reconstitution of, 95
Nucleoside diphosphate reductase, 31–32, 57

Nucleus, uptake of proteins by, 122

O

Oncogenic virus, 137
Onion root cells, 61
Oocytes, cytoplasmic extracts of, 67
Orotate phosphoribosyltransferase, 128
Oxytricha, 131

P

Pancreas, 48
Paramecium, 22
    macronuclear G_1, 49
Paramecium aurelia, 62
Parenchymal cells, 47
Pea root tips, 43
Pelomyxa, 60
Phosphate transport, 114
Phosphodiesterase, 109–110, 112, 114
Phosphokinases, 121
Phospholipid, 103
Phospholipid vesicles, cAMP-containing, 112
Physarum, 83, 120–121, 128
    acceleration of cycle in, 121
    DNA replication in, 72
    G_2 delay of nuclei of, 88
    protein kinase in, 118
    replication of ribosomal cistrons in, 82
    synthesis of RNA in, 126
Physarum polycephalum, 49
    cell cycle of, 20
    synthesis of nucleolus-associated DNA, 21
Phytohemagglutinin, 114, 118
    effect on levels of cGMP, 115
Pinus pinea, 49
Plant embryos, DNA content of, 48
Plasma membrane, 102, 117
    carbohydrate, amount of, 103–104
    chemical changes of, 103
    morphological change in, 103
    nutrients, transport of, 114
    protein, amount of, 103–104
Platy-swordtail hybrids, 136
Polyoma-transformed cells
    BHK, 115
    mouse, 137
Polyoma virus, 115
    temperature-sensitive mutants of, 136

Polysomes
    disappearance during mitosis, 93
    histone, 119
Polytene chromosome(s), 129
    formation, 82
    replicating unit in, 74
    replication origins in, 78
Polyteny, 52, 131
Pronase, release from density-dependent
    inhibition with, 46
Prostaglandin E_1, 114
Protein(s)
    release from nucleus, 91
    return to the nucleus after mitosis, 94
    reversible shift of, 94
Protein kinases, 118
Protein synthesis
    amino acid deprivation and rate of, 55
    in G_1, 53
Puromycin, 53
Py3T3 cells, 112

R

R point, 46
Rabbit euploid cells, 80
Rat kangaroo cells, replication of ribosomal
    cistrons in, 82
Regulation, dominance of, 137
Regulation of cell reproduction, loss of, 135
Regulatory genes
    concept of for cell reproduction, 135
    mutation in, 138
Repetitious DNA, replication of, 81
Replicating units (replicons)
    bank of, 76
        initiation of, 80
    length of, 75
    number of in eukaryotic cells, 74
    number of in mammalian cells, 75
    per chromosome, 75
    size in *Drosophila,* 78
    sizes of, changes in, 76
Replication, change in the number of origins
    of, 76
Replication band, 68, 71
Replication fork(s), 68, 72
    number of in HeLa, 74
    rate of travel of, 75
    rate of traverse of, 83
Replication origins, 72

Replication pattern, of chromosomes, 80
Replicon families, minimum number of, 75
Resynchronization, of mitotically selected
  cells, 33
Retroactive synchronization, 80
Ribonuclease, 128
Ribosomal precursor RNA, 95
Ribosomal RNA synthesis, blocking in G_1,
  53
Ribosomes, incompetence of during mitosis,
  93
RNA
  low molecular weight, 125
  reversible shift of nuclear, 94
RNA synthesis
  and the cell cycle, 124
  cessation during mitosis, 92
  rate of, 124–126
mRNA
  carry over through mitosis, 93
  half-life of, 93
  viral, 116
mRNA synthesis, requirement for, 124
Root meristem, 47
Rous sarcoma virus, 116

S

S period
  definition of, 3
  determination of, 30
  GC to AT shift during, 80
  initiation of, 60
  length in cells of different ploidies, 85
  length of, 29
    relation with genome size, 86
    variation of, 30
  protein synthesis, requirements, 83
  ribosomal RNA synthesis, inhibition, 90
  RNA synthesis, requirements, 83
  synchrony of entry into, 37–38
Saccharomyces cerevisiae, 131
  initiation of DNA synthesis in, 53
  mitochondrial DNA synthesis in, 64
Saccharomyces lactis, mitochondrial DNA
  synthesis in, 64
Salivary gland cells, stimulation of, 122
Salmonella, 136
Satellite DNA, 76
  timing of replication of, 81
Schizosaccharomyces pombe, 8, 49, 127

Sea squirt, 130
Sea urchin eggs, 19
Sea urchin embryos
  cAMP in, 108–109
  cleavage stages of, 51
Separation of cells by size, 21
Serum deprivation, 134
Skeletal muscle cells, 48
Slime mold, 49, 64, 80, 128
  DNA synthesis in, 60
  mitochondrial, 63
  synchrony, 79
  G_1 period, lack of, 50
  protein synthesis, inhibition of, 50
  thymidine kinase in, 55
Smooth muscle, 48
Snail embryos, 51
Splenic lymphocytes, 115
Stentor, 64
  cytoplasmic/nuclear ratio, 42
  DNA synthesis, initiation of, 42
  macronuclear G_1, 49
Succinic dehydrogenase, 128
Surface area, 102
Surface change, 117
Surface morphology, changes in, 98
SV3T3 cells, 115
  growth curve for, 102
SV40, 115–116
  activation of cellular DNA synthesis by,
    136
SV40-transformed human cell, 137
Synchrony
  decay in, 37
  deterioration of, 28–30
  experimentally derived, 21
  induced, 31
  loss of, 22–23, 38
  natural, 19
  of mitotically selected cells, 28
  retroactive, 29–30
  selection, 21
  two step, 33

T

T antigen, location during mitosis, 95
3T3 cells, 102, 114–115
  cAMP levels in, 113
3T6 cells, 125
  ghost, 126

release from growth arrest, 35
Temperature-sensitive mutant(s), 132–133,
    138
  of Rous sarcoma virus, 117
Temperature-sensitive mutations, 131
Tetrahymena, 22, 62
  arrest in macronuclear G_1, 45
  cell cycle growth in, 14
  cell cycle of, 14–15
  deoxynucleoside triphosphate pools in, 57
  DNA synthesis, synchrony of, 31
  rDNA, replication of, 82
  generation times, distribution of, 23
  macronuclear G_1, 49
  micronucleus of, 49
  respiration rate during the cell cycle in,
    14–15
  synchronization of, 31
  synchrony, loss of, 23
  volume increase in, 14–15
Tetrahymena pyriformis, cell cycle of, 63
Theophylline, 112, 114, 117
Thymidine kinase, 113, 129
  and the cell cycle, 55–57
Thymidine nucleotides, pool of, 31, 54
Thymidylate synthetase, 31–32
  in G_1, 56
Thymus, 43
Tongue, 46
Transformation, 135
Transformed state, recessiveness of, 138
Translation, inhibition of during mitosis, 93
Triturus
  embryonic cells of, 76
  meiosis in, 76
  S period in, 76
  somatic cells of, 76
Tryptophan
  analogues of, 84
  deprivation of, 54
TTP, formation of, 32
Tumor malignancy, suppression of, 137
Tyrosine aminotransferase, 128

U

Uridine nucleotide, reduction of, 57
Uridine transport, 114
Urostyla, 60

V

V79 line, 38, 51
  cells, 125
Vaccinia virus, replication of, 94
Velocity sedimentation of cells, 21
Vicia faba, 49

W

WI-38 cells, 45, 122–123

X

X chromosome
  of the grasshopper, 81
  replication of heterochromatic, 81
Xenopus, 67
  haploid cells of, 85
  nuclei, injection of into eggs of, 65
Xenopus embryos, 51

Y

Y chromosome
  replication during spermatogenesis, 81
  replication in somatic cells, 81
Yeast, 49, 84, 127
  cell cycle, 6, 53, 131–132
    carbohydrate synthesis in, 10
    growth in, 8
    protein synthesis, in, 10
    regulation of, 5
    RNA synthesis, in, 10
    sections of, 9
  DNA synthesis, initiation of, in, 53
  growth in dry mass, 8–9
  growth of a synchronous population of,
    21
  mating factor, 5
  mating type(s), 5, 132
  mitochondrial DNA synthesis in, 63
  pool size in, 10
  scanning electron micrograph, 9
  temperature sensitive mutants of, 53
  volume measurements, 8–9

Z

Zea mays, 47